BUSINESS AND TECHNICAL COMMUNICATION

•

A Bibliography, 1975-1985

by

DEBRA L. HULL

The Scarecrow Press, Inc.
Metuchen, N.J., & London
1987

Library of Congress Cataloging-in-Publication Data

Hull, Debra L., 1958–
    Business and technical communication.

    Bibliography: p.
    Includes indexes.
    1. Communication in management--Bibliography.
2. Business writing--Bibliography. 3. Communication
of technical information--Bibliography. 4. Technical
writing--Bibliography. I. Title.
Z7164.C81H7 1987 [HF5718] 016.00151   87-4749
ISBN 0-8108-1971-6

FOR ELIZABETH TEBEAUX

Doctrine and life, colours and light, in one
    When they combine and mingle, bring
A strong regard and awe: but speech alone
    Doth vanish like a flaring thing,
    And in the ear, not conscience ring.

                        --George Herbert

• CONTENTS •

vi

• ACKNOWLEDGMENTS •

I wish to express my gratitude to Norman S. Grabo for suggesting that I publish this bibliography and to both Claude Gibson and Guy Bailey for making sure that I followed that suggestion. I would like to thank J.J. and Sharon Dent, Rick Evans, Gwendolyn Gong, and Debra Hower for the encouragement they gave. I am greatly in debt to Karen Forrest and, particularly, Diana Luna for the many hours they devoted to help type the manuscript. I am especially grateful to my friend Shirley Bovey for listening and understanding.

Finally, I reserve special thanks for my parents for always believing in me.

Debra L. Hull
College Station, Texas

• INTRODUCTION •

The growth in the amount of data that must be converted into information is staggering; new means for transmitting information are rapidly developing; and business organizations are moving away from hierarchical structures to the more interactive matrix structures. These changes demand that the teachers of business and technical communication constantly reexamine their definitions of business and technical communication and restructure their courses so that they can prepare students for solving new communication problems that are arising with these changes.

This bibliography, covering the period from 1975 to 1985, is intended to be an aid to business and technical communication teachers by providing easy access to literature concerning teaching in this field. This collection allows teachers to examine the changes that have taken place in the last ten years and to project what is to come in the future.

To compile the bibliography, I began with a search for articles listed in major education indexes: The Current Index to Journals in Education, ERIC Index, and the Education Index. I then expanded my search to include periodicals not covered in the indexes: Journal of Technical Writing and Communication, The Journal of Business Communication, IEEE Transactions on Professional Communications, and Technical Communication. Because countless books on business and technical communication exist and because the "Bibliographies" section includes bibliographies on books, I limited my search for books to those that I could actually examine at the Texas A&M University library and to those listed in the most recent issue of Books in Print.

In my collection, I eliminated articles on the teaching of business and technical communication in such specialized fields as law and medicine since the language of these fields is too specialized to include in a general business and technical communication classroom. I also eliminated articles on programs to teach business and technical communication to foreign students since these programs also differ from the general classroom.

The bibliography is organized according to areas that are important to the teacher of business and technical communication. Part I provides insight into the nature of the business and technical communication classroom. As can be seen by the number of articles listed, many have tried to define what is meant by the phrase "business and technical communication." These definitions have and still do vary greatly; therefore, a consensus as to what should be taught in the business and technical communication classroom has not been reached. Arising from this lack of consensus are numerous teaching programs--each with its own advantages and disadvantages.

No matter what definitions teachers accept or what programs teachers follow, they must choose teaching methods and grading techniques. Again, however, choices of methods and techniques abound as educators try to determine what roles as instructors and evaluators the teachers have in the classroom.

Once programs are established, course content must be designed. Thus, Part II of the bibliography examines possible contents of the business and technical communication classroom.

So that they can become effective communicators, students must first be taught how to analyze their audiences and how to make stylistic choices that are appropriate to the defined audiences. Students also need to consider how to present the material--through either written documents or oral presentations. In both methods of presentation, graphics are a consideration and, if used, should be incorporated correctly and effectively. Finally, because presentations should contain accurate business and technical information, students need to understand the importance of and methods for thorough library research.

Even after programs and course designs are established,

the instructors must constantly update to prepare students for the world beyond the classroom. Part III of the bibliography explores how the business and technical communication classroom is affected by the technological advances that rapidly are changing the nature of communication in business and industry. Educators are only beginning to define the changes arising from the use of computers and from developments in telecommunications and teleconferencing and to devise means for preparing students to meet, define, and solve new communication problems.

PART I:

THE BUSINESS AND TECHNICAL
COMMUNICATION CLASSROOM

1. SEEKING A DEFINITION •

1.1 Influences from the Past

1. Baker, Christopher. "Francis Bacon and the Technology
    of Style." Technical Writing Teacher 10 (1983): 118-123.
    States that content is not the only feature which dis-
    tinguishes scientific communication. Suggests that study-
    ing the works of Francis Bacon can help determine the
    "origins of modern technical style" (118) because "his
    technology of style ... is not merely a reaction to older
    stylistic schools, but springs directly from his philosophy
    of science" (120).

2. Bareisich, Michael J. "The Relativity of Communication:
    Albert Einstein As Technical Writer." Journal of Tech-
    nical Writing and Communication 10 (1980): 125-132.
    Uses Albert Einstein's writings to teach audience
    awareness. Shows how Einstein wrote about his theory
    of relativity to three different audiences.

3. Brockmann, R. John. "Bibliography of Articles on the
    History of Technical Writing." Journal of Technical Writ-
    ing and Communication 13 (1983): 155-65.
    Lists and annotates 36 articles concerning the history
    of technical writing.

4. Broughton, Bradford B. "'No Man Is Allowed to Spell
    Ill': Modern Communication Advice from an Eighteenth
    Century Expert." Journal of Technical Writing and Com-
    munication 15 (1985): 157-161.
    Shows that Lord Chesterfield's writings on how to ef-
    fectively communicate in the 18th century contains valu-
    able suggestions for the modern day communicator.

5. Conners, Robert J. "The Rise of Technical Writing

3

Instruction in America." Journal of Technical Writing and Communication 12 (1982):   329-52.

Examines the development of technical writing instruction from 1900 to 1980.

6.   Dobrin, David N.   "Is Technical Writing Particularly Objective?" College English 47 (1985):   237-51.

Examines the belief that technical writing is objective. Explains the problems that come from this belief.

7.   Gresham, Stephen L.   "Benjamin Franklin's Contributions to the Development of Technical Communication." Journal of Technical Writing and Communication 7 (1977):   5-13.

Explains how Franklin's style is a good model for technical writers.

8.   Limaye, Mohan R.   "The Syntax of Persuasion: Two Business Letters of Request." The Journal of Business Communication 20.2 (1983):   17-30.

Analyzes two letters of request written in the 1590s-- "a model letter of request from The English Secretary written by Angel Day and a letter requesting a loan written by Richard Quiney to William Shakespeare" (18)--to determine if current guidelines for writing requests are put to actual use in these letters.

9.   Lipson, Carol S.   "Francis Bacon and Plain Scientific Prose: A Reexamination." Journal of Technical Writing and Communication 15 (1985):   143-155.

Argues that Bacon's theory of rhetoric commonly viewed as promoting the use of plain style in scientific writing does not actually reflect this view.

10.   Locker, Kitty O.   " 'Sir, This Will Never Do':   Model Dunning Letters, 1592-1873." The Journal of Business Communication 22.2 (1985):   39-45.

Lists and describes three classes of "dunning" letters that appear in English letter writing between 1568 and 1897:   apologetic, vituperative, and businesslike.   Explains that by the nineteenth century "both apologetic and vituperative dunning letters were being replaced by more businesslike letters" (43).

11.   Markel, Michael M.   "The Rhetorical Principles of Sir

Thomas Browne." Technical Communication 26.1 (1979):
8-9.
Explains what technical communicators can learn from
the writing of Sir Thomas Browne.

12. Matalene, H. W. "The Interaction of the Anciene Régime;
or, Why Does Anybody Ever Bother to Listen to Anybody
Else?" College English 46 (1984): 22-31.
Describes what constituted Europe's ancien régime
and explains how the ancien régime still affects us today.

13. Minor, Dennis E. "Newspeak, 1984, and Technical Writing." Journal of Technical Writing and Communication
15 (1985): 365-72.
Draws from Orwell's works to show how language can
easily be turned into propaganda. Reminds technical
writers of their responsibilities.

14. Moran, Michael G. "Joseph Priestley, William Duncan and
Analytic Arrangement in 18th Century Scientific Discourse."
Journal of Technical Writing and Communication 14 (1984):
207-215.
Examines the use of inductive reasoning in the works
of Priestley and Duncan. Presents some of the "rhetorical advantages" of inductive presentation of written material.

15. _____. "Writing Business Correspondence Using the
Persuasive Sequence." The ABCA Bulletin 47.2 (1984):
24-27.
Explains how Joseph Priestley's method of arranging
information for persuasive purposes can be used by writers as invention and arrangement techniques.

16. Ohmann, Richard. "Literacy, Technology, and Monopoly
Capital." College English 47 (1985): 675-89.
Examines the relationship between literacy and tech-
nology.

17. Post, Robert. "My Favorite Assignment." The ABCA
Bulletin 45.4 (1982): 26-30.
Describes an assignment using George Orwell's "Pol-
itics and The English Language."

18. Rosner, Mary. "Style and Audience in Technical Writing:

Advice from the Early Texts." Technical Writing Teacher
11 (1983): 38-45.
Compares discussions of style and audience found in
seventeen early technical writing texts to discussions
found in three modern texts. Concludes that "advice"
given in the early texts differs little from that given
in the modern texts. Offers suggestions for improving
the instruction given in technical writing textbooks.

19. Schwartz, Mimi. "Response to Writing: A College-Wide
    Perspective." College English 46 (1984): 55-62.
    Describes a study conducted to determine what value
    various disciplines at the college level place on writing.

20. Scott, Geoffrey L. "The Scientific Poetry of Erasmus
    Darwin." Technical Communication 29.3 (1982): 16-20.
    Shows what the technical writers can learn by study-
    ing the poetry of Erasmus Darwin.

21. Whalen, Tim. "A History of Specifications: Technical
    Writing in Perspective." Journal of Technical Writing
    and Communication 15 (1985): 235-45.
    Describes the development of specification writing.

22. Whitburn, Merrill D. "The Past and the Future of Scien-
    tific and Technical Writing." Journal of Technical Writ-
    ing and Communication 7 (1977): 143-49.
    Shows what influences the rise of science in the
    seventeenth century had on present day scientific and
    technical writing.

23. Whitburn, Merrill D.; Marijane Davis; Sharon Higgins;
    Linsey Oates; and Karen Spurgeon. "The Plain Style
    in Scientific and Technical Writing." Journal of Techni-
    cal Writing and Communication 8 (1978): 349-57.
    Discusses changes in scientific style that paralleled
    the rise of science in the seventeenth century.

24. Woff, Luella M. "A Brief History of the Art of Dictamen:
    Medieval Origins of Business Letter Writing." Journal
    of Business Communication 16.2 (1979): 3-11.
    Discusses the development of letter writing.

1.2 Determining the Nature of Technical Communication

25.   Aldrich, Pearl G. "Adult Writers: Some Reasons for
      Ineffective Writing on the Job." College Composition
      and Communication 33 (1982): 284-87.
            Describes a study of people with scientific, technical,
      and legal degrees to determine their writing ability.

26.   Bowman, Joel P. "From Chaos to K-Mart: Beyond the
      You-attitude." The ABCA Bulletin 47.4 (1984): 16-20.
            Examines the traditional communication hierarchy of
      an organization and describes changes that are affecting
      this hierarchy.

27.   Bowes, John E., and Keith R. Stamm. "Science Writing
      Techniques and Methods: What the Research Tells Us."
      Association for Education in Journalism Conference,
      Seattle. 13-16 August 1978.
            Describes what kinds of research exist concerning
      science writing.

28.   Bromage, M. C. "Bridging the Corporate Communica-
      tions Gap." S.A.M. Advanced Management Journal 41.1
      (1976): 44-51.
            Examines differences between writing done by mana-
      gers and writing done by technicians.

29.   Cochran, Daniel S., and Janet A. Dolan. "Qualitative
      Research: An Alternative to Quantitative Research in
      Communication." The Journal of Business Communica-
      tion 21.4 (1984): 25-32.
            Speculates the causes for the lack of communication
      research. Suggests the use of "qualitative" research
      techniques instead of "quantitative" research techniques.

30.   Covington, David H., and Clifford M. Krowne. "A Sur-
      vey of Technical Communication Students: Attitudes,
      Skills, and Aspirations." Journal of Technical Writing
      and Communication 13 (1983): 205-19.
            Conducted a survey of 274 technical communication
      students at North Carolina State University to discover
      students' views of technical communication.

31.   Crowe, Ronald. "Keep It Simple, Keep It Clear."
      Writer's Digest 55 (1975): 14-20.
            Discusses what constitutes technical writing and how
      to produce good technical writing.

32. Curley, Stephen J. "Is the Teaching of Technical Writ-
     ing Really Relevant?" Journal of Technical Writing and
     Communication 7 (1977):   309-24.
         Examines arguments for and against the relevance of
     teaching technical writing.

33. Daniel, Carter A. "Carter Daniel Replies." The Journal
     of Business Communication 21.3 (1984):   31-2.
         Responds to others' comments on his 1983 article
     " 'Remembering Our Charter: Business Communication at
     the Crossroads.' " Restates his main conclusions.

34. _____. "Remembering Our Charter: Business Com-
     munication at the Crossroads." The Journal of Business
     Communication 20.3 (1983):   3-11.
         Examines a perceived "decline" of the field of business
     communication. Suggests ways to curtail this decline.

35. Dorazio, Patricia. "Impressions of the Technical Writer:
     A Master of Many Communication Roles." The ABCA
     Bulletin 48.2 (1985):   16-20.
         Presents the author's views of the multifaceted roles
     she assumes as a technical writer for IBM, Kingston,
     New York.

36. Durfee, Patricia Bernadt. "Responding to Industry:
     Writing in a High Tech World." Improving College and
     University Teaching 32.4 (1984):   180-84.
         Discusses a study of what communication skills are
     required of technicians in industry.

37. Evand, William H. "Who Is Qualified to Teach Business
     Writing and Technical Writing?" Conference on CCC,
     Washington.   13-15 March 1980.
         Describes a study to determine who is teaching tech-
     nical writing and what they are teaching.

38. Feinberg, Susan G. "A Questionnaire to Evaluate Stu-
     dent Attitude." Technical Writing Teacher 11 (1983):
     46-51.
         Discusses a study conducted to determine "changes
     in student attitude toward technical communication" (46).
     Concludes that such a study does accurately measure
     changes in attitude. Provides a sample questionnaire
     that teachers can use to conduct their own studies.

39. Fine, Marlene G. "Writing Skills in Business: Needs, Training Strategies, and Double Binds." The ABCA Bulletin 46.2 (1983): 12-16.

    Discusses the writing needs of corporate employees. Explains how the corporate environment hinders development of good writing skills.

40. Fontenot, Karen, and Steven Golen. "Business Communication and Journalism: Sharing a Common Goal of Effective Communication." The ABCA Bulletin 46.3 (1983): 12.

    Presents similarities between journalism and business communication.

41. Girill, T. R. "Technical Communication and Journalism." Technical Communication 31.3 (1984): 40.

    Discusses areas of concern that are shared by both journalists and technical communicators.

42. _____. "Technical Communication and Philosophy." Technical Communication 31.1 (1984): 37.

    Explains the importance of " 'philosophical' skills" to technical communicators.

43. Glassman, Myron, and Thomas E. Pinilli. "Scientific Inquiry and Technical Communication: An Introduction to the Research Process." Technical Communication 32.4 (1985): 8-13.

    Stresses the importance of technical communicators' being familiar with research processes. Discusses various types of research methodologies and examines steps in the research process.

44. Goldberg, Jay J. "A Survey of Scholarly Works in Technical Writing." Technical Communication 22.1 (1975): 5-8.

    Examines scholarly works on technical writing that were published from 1965 to 1973.

45. Goodell, Rae. "Should Scientists Be Involved in Teaching Science Writing and If So, How?" Association for Education in Journalism Conference, Seattle. 13-16 August 1978.

    Examines the role of scientists in the education of technical writers.

46.   Gould, Jay R. "Concerns of the Technical Writer: A
      Panel Discussion." Journal of Technical Writing and
      Communication 6 (1976):   163-67.
          Examines concerns expressed in a panel discussion
      at the 1975 annual technical writers' Institute of Rens-
      selaer Polytechnic Institute.

47.   Green, Marcus M., and Timothy D. Nolan. "A System-
      atic Analysis of the Technical Communicator's Job: A
      Guide for Educators." Technical Communication 31.4
      (1984):   9-12.
          Discusses a survey conducted to determine what is
      required of technical communicators on the job.

48.   Gresham, Steve, and Carey Kaltenback. "Readings in
      Technical Writing: An Orientation Workshop." Techni-
      cal Writing Teacher 2 (1975):   4-8.
          Describes a two-week workshop which consists of
      students reading and discussing articles on technical
      writing. Believes that this workshop, conducted at the
      beginning of the semester, helps students answer for
      themselves the question, "What is technical writing?"
      (4).

49.   Hamlin, Donna M., and Craig Harkins. "A Model for
      Technical Communication." Journal of Technical Writing
      and Communication 13 (1983):   57-81.
          Draws from a variety of sources to present a new
      model for technical communication.

50.   Harris, John S. "On Expanding the Definition of Tech-
      nical Writing." Journal of Technical Writing and Com-
      munication 8 (1978):   133-38.
          Examines how the definition of technical writing is
      changing.

51.   Hays, Robert. "Political Realities in Reader/Situation
      Analysis." Technical Communication 31.1 (1984):   16-20.
          Discusses the "politics" that affect writers and
      readers of documents. Suggests ways for handling "po-
      litical" situations.

52.   Herndl, Carl G. "Hierarchies of Audiences and Texts."
      Conference on CCC 35th NY, NY March 29-31 1984.
          Describes a study to determine writing problems of
      professionals on the job.

53. Hertz, Vivienne. "Cognitive-Field Implications for the Teaching of Technical Writing." International Conference on Improving University Teaching, Newcastle-upon-Tyne, England. 8-11 June 1977.
    Describes what cognitive-field psychology has to offer the technical writing classroom.

54. Hische, Gerard. "Technical Writing Is Different." IEEE --Transactions on Professional Communications PC-20 (1977): 15-17.
    Describes considerations of technical writers.

55. Kelly, Kathleen. "Business Writing and the Humanities: Having It Both Ways." Pennsylvania State Conference on Rhetoric and Composition, University Park, Pennsylvania July 7, 1982.
    Explains how to design a business writing course that is both humanistic and technical.

56. Kotler, Janet. "On I. A. Richards--And Some Other Things." The ABCA Bulletin 48.1 (1985): 2-4.
    Argues that communication research should focus on determining what constitutes "successful (and unsuccessful) communication" (4).

57. Kowal, John Paul. "Training Tomorrow's Writers." Journal of Technical Writing and Communication 5 (1975): 181-85.
    Examines the importance of effective communication to technicians.

58. Limaye, Mohan R. "Redefining Business and Technical Writing by Means of a Six-Factored Communication Model." Journal of Technical Writing and Communication 13 (1983): 331-40.
    Uses Jakobson's model of verbal communication to develop a definition of business and technical writing.

59. Locker, Kitty O. "What Do Writers in Industry Write?" Technical Writing Teacher 9 (1982): 122-27.
    Examines the kinds of writing done outside academia.

60. Logan, Robert A. "Secularization and Popularization." Association for Education in Journalism Conference, Seattle, 13-16 August 1978.

Discusses the issue of what scientific information
should be presented to the public and how.

61. Ozake, Shigeru. "Business English from a Human Point
of View." The Journal of Business Communication 12.2
(1975): 27-31.
Argues that emphasis should be placed not only on
the linguistic element of business communication but
also on the human element in business communication.

62. Palumbo, Roberta M. "Professional Writing for the Liberal Arts Student: An Untapped Resource." Regional
Meeting of ABCA Hammond, Louisana. 5-7 April 1984.
Explains how students in a professional writing course
can learn about the use of rhetoric in professional writing.

63. Pauly, John. "The Case for a New Model of Business
Communication." The Journal of Business Communication
14.4 (1977): 11-23.
Presents a new model for the definition of communication.

64. Penrose, John M. "A Survey of the Perceived Importance
of Business Communication and other Business--Related
Abilities." The Journal of Business Communication 13.2
(1976): 17-24.
Describes a study to determine the attitudes that
those in business have toward business communication.

65. Persing, Bobbye Sorrels. "Search and Re-search for
Solutions to Communication Problems." The Journal of
Business Communication 16.2 (1979): 13-25.
Discusses the scope and aims of communication re-
search.

66. Peterson, Bruce T. "Conceptual Patterns in Industrial
and Academic Discourse." Journal of Technical Writing
and Communication 14 (1984): 95-107.
Describes a study that draws from the works of Mof-
fett and Britton to establish criteria for determing if
technical writing does no more than present factual information.

67. _____. "Technical Writing, Revision, and Language

Communities." Conference on CCC, 35th New York, New York. 29-31 March 1984.

Describes a study to determine the writing process of those on the job.

68. "Responses to Carter A. Daniel's Article 'Remembering Our Charter: Business Communication at the Cross-roads,' Summer 1983." The Journal of Business Communication 21.3 (1984): 17-31.

Presents numerous responses made to Daniel's article.

69. Rothmel, Steven Zachary. "Technical and Creative Writing: Common Process, Common Goals." Conference on CCC, Washington, D.C., 13-15 March 1980.

Examines similarities between creative writing and technical writing.

70. Rubens, Philip M. "The Writer's Mind: Ethics in the Teaching of Technical Writing." Meeting of the New York College English Association, Saratoga Springs, New York, 3-4 October 1980.

Describes three ethical issues of concern to technical writers.

71. Sager, Eric. "Operational Definition." The Journal of Business Communication 14.1 (1976): 22-26.

Discusses the usefulness of operational definition to research.

72. Samuels, Marilyn Schauer. "Technical Writing and the Recreation of Reality." Journal of Technical Writing and Communication 15 (1985): 3-13.

Discusses the creative nature of technical writing.

73. Sawyer, Thomas M. "Mix Student Disciplines in Technical Writing Courses." Conference on CCC, Denver, 30 March-1 April 1978.

Examines the advantages of teaching a technical writing class that consists of students from different majors. Discusses what should be taught in such a course.

74.

Schall, Maryan. "Profile of the Technical Communication Student at the University of Minnesota." Technical Communication 23.3 (1976): 2-6.

Describes a survey conducted to determine the characteristics of technical communication students.

75.   Selzer, Jack. "Some Differences Between Journalism
      and Business Writing." ABCA Bulletin 46.3 (1983):   8-
      10.
          Finds that journalism courses have a lot to offer
      business writing students but points out that business
      writers must also perform writing tasks that are not
      within the realm of journalism

76.   Sides, Charles H. "Technical Writing: Implications for
      Compositional Skills Development." National Council of
      Teachers of English Conference, San Francisco, 22-24
      November 1979.
          Attempts to define technical writing.

77.   Smeltzer, Larry R.; Janice Glab; and Steven Golen.
      "Managerial Communication: The Merging of Business
      Communication, Organizational Communication, and
      Management." The Journal of Business Communication
      20.4 (1983):   72-78.
          Discusses the rise of a new discipline, managerial
      communication, that is being formed by the "merging"
      business communication, organizational communication,
      and management.

78.   Sparrow, W. Keats. "Six Myths About 'Writing for
      Business and Industry.'" Technical Writing Teacher
      3 (1976):   49-59.
          Dispels six "myths" about business and technical
      writing that often lead to misunderstandings concerning
      both the teaching and practicing of business and techni-
      cal writing.

79.   _____. "Technical Writing as a Liberal Arts Skill."
      College English Association Conference, Dearborn, Michi-
      gan, 10-12 April 1980.
          Describes how technical writing should be taught as
      a liberal arts course.

80.   Sullivan, Dale. "Connections with the Liberal Arts and
      Industry: Attempts to Legitimize the Profession of Teach-
      ing Technical Writing." Conference on CCC 36th Min-
      neapolis, Minnesota, 21-23 March 1985.
          Discusses issues concerning the definition of techni-
      cal writing. Includes a bibliography.

81. Varner, Iris I., and Carson H. Varner. "Legal Issues in Business Communications." ABCA Bulletin 46.3 (1983): 14-19.
    Discusses in detail the role that legal issues have in business communication.

82. Weber, Barbara. "Technical Writing Skills: A Question of Aptitude or Interest?" Journal of Technical Writing and Communication 15 (1985): 63-68.
    Explains why professional writers often are better communicators than technicians.

83. Wicclair, Mark R., and David K. Farkas. "Ethical Reasoning in Technical Communication: A Practical Framework." Technical Communication 31.2 (1984): 15-19.
    Examines ethical problems that technical communicators sometime face. Provides some case analyses.

1.3 Technical Writing and Composition Instruction

84. Brockmann, R. John. "What's Different About Teaching Technical Writing." Technical Writing Teacher 10 (1983): 174-181.
    Discusses six areas of contrast to distinguish the teaching of technical writing from the teaching of composition and/or the teaching of literature.

85. Bruckman, C. G. "What Is in a Name? Freshman English or Technical Writing." Journal of Technical Writing and Communication 6 (1976): 187-93.
    Addresses the topic of whether or not technical writing should replace freshman English in answer to contradictory arguments raised by W. Earl Britton and Merril D. Whitburn. Describes a course offered at the University of the Witwaters, which serves as a compromise by merging information provided in freshman English and information provided in technical writing.

86. Collins, Terence. "What's New in Freshman Comp? Gleanings for the Business Writing Teacher." ABCA Bulletin 44.1 (1981): 10-12.
    Finds similarities between teaching business writing and teaching freshman composition. Explains how research

on the composing process, rhetorical modes, and classroom peer groups is important for business writing
teachers.

87.   Dietrich, Julia C.   "The Common Ground of Freshman
      Rhetoric and Technical Writing."   Technical Writing
      Teacher 6 (1979):   92-94.
          Describes an experimental freshman English course
      taught at Case Western Reserve University. Explains
      how the course tries to reach "the common ground of
      technical writing and freshman rhetoric"(93).

88.   Hursey, Richard C.   "Abandon the Barricades."   Teach-
      ing English in the Two-Year College 6 (1979):   21-23.
          Suggests that in order to be relevant to students,
      freshman English courses should emphasize contemporary
      readings and technical writing.

89.   Pauly, John J.   "Journalism and Business Writing:
      Roundtable."   ABCA Bulletin 46.3 (1983):   6-8.
          Argues that journalism and business writing classes
      better prepare students to write than freshman composi-
      tion courses.

90.   Raisman, Neal A.   "Technical Writing Reduces Writing
      Anxiety: Results of a Two-Year Study."   Technical
      Writing Teacher 11 (1984):   145-56.
          Describes a two-year study that "found that students
      taught technical writing in their freshman composition
      courses became less fearful of instruction in writing than
      students taught literary--essay or rhetorically-based
      writing" (145). Gives reasons for why students felt less
      anxiety when "learning to write through instruction in
      technical writing" (154).

91.   Tebeaux, Elizabeth.   "Technical Writing Is Not Enough."
      Engineering Education 70.7 (1980):   741-43.
          Describes the relationship between freshman composi-
      tion and technical writing at the College of Technology
      of the University of Houston.

92.   Tichy, Henrietta.   "To the New Teacher of Technical
      Writing."   Journal of Technical Writing and Communica-
      tion 7 (1977):   251-59.
          Examines similarities and differences in teaching college
      composition and teaching technical writing.

93.   Whitburn, Merrill D. "Against Substituting Technical
      Writing for Freshman English." Journal of Technical
      Writing and Communication 5 (1975): 47-51.
          Argues against W. Earl Britton's proposal to replace
      freshman English with technical writing. Examines the
      different intents (focuses) of instruction in freshman
      English and technical writing to show the merits of both
      courses. States that because focuses of the courses
      are different--i.e., technical writing builds upon the
      body of knowledge students obtain in freshman composi-
      tion--then technical writing cannot replace freshman
      English.

1.4 Technical Writing and Literary Studies

94.   Anderson, Paul V. "Background and Resources for the
      New Teachers of Technical Writing." Journal of Techni-
      cal Writing and Communication 7 (1977): 223-33.
          Discusses the move from teaching literature to teach-
      ing technical writing. Provides some resources that
      teachers can use to better prepare for the teaching of
      technical writing.

95.   Barnum, Carol M. "The Metamorphosis of the English
      Teacher: From Lit Jock to Tech Writer." Technical
      Writing Teacher 9 (1981): 25-28.
          Discusses the reasons for the increased "popularity" of
      technical writing courses. Examines the role of English
      teachers in the teaching of technical writing and offers
      suggestions as to how teachers with backgrounds in
      literature can better prepare for teaching technical writ-
      ing.

96.   Edens, Bert. "Readability and Creativity in Technical
      Writing." Journal of Technical Writing and Communica-
      tion 10 (1980): 329-36.
          Discusses the differences between literary writing
      and technical writing but then shows how the technical
      writer can learn about creativity from the study of liter-
      ature and can apply this creativity to technical writing.
      Suggests that the study of essays in secondary schools
      is one way to begin training creative technical writers.

97.   Gould, Jay R. "Bringing Teachers of Technical Writing

and Teachers of Literature Closer Together." Journal
of Technical Writing and Communication 9 (1979): 173-
83.
    Examines ways that technical writing teachers and
literature teachers can learn from each other and discusses common goals that these teachers have.   Suggests that "any discussion on writing should provide
a forum for both kinds of teachers" (182).

98. Halloran, S. Michael. "What Every Department Chair
     Should Know about Scholarship in Technical Communica-
     tion." ADE Bulletin 79 (1984):   43-45.
     Discusses differences in literary scholarship and
technical communication scholarship.

99. Kroiter, Harry P., and Elizabeth Tebeaux. "Bringing
     Literature Teachers and Writing Teachers Closer To-
     gether." ADE Bulletin 78 (1984):   28-32.
     Examines issues that divide literature teachers and
writing teachers.   Suggests ways that literature teachers
and writing teachers can work together.

100. Larson, Richard. "How Not to Be a 'Retread.'" Tech-
     nical Writing Teacher 9 (1982):   84-85.
     Suggests ways that literature teachers can prepare
for teaching composition and technical writing.

101 McNallie, Robin. "'Leaving the Old': Reflections of a
     Middle-Aged English Teacher on His First Year in the
     Technical Writing Classroom." Technical Writing Teacher
     10 (1982):   50-53.
     Discusses the author's experience of moving from
teaching literature to teaching technical writing.

102. Orth, Mel F. "Objectives in Technical Writing." Tech-
     nical Writing Teacher 8 (1981):   51-53.
     Argues that the objectives in technical writing are
not only "utilitarian" but also "humanistic" (50). Shows
what makes technical writing humanistic.

103. Pickett, Nell Ann. "From Researching Colloquialism as
     a Style in the First-Person Narrator Fiction of Eudora
     Welty to Explaining Why a Mule Can't Reproduce: or,
     The Reduction of an English Teacher." Journal of
     Technical Writing and Communication 9 (1979):   51-57.

Examines the similarities and differences between
teaching literature and teaching technical writing by
disucssing the needs of students enrolled in the community college where the author teaches. States that
teachers and curriculums should first meet the needs
of the students. Finds that students in community colleges tend to be more interested in practical rather
than literary topics.

104. Rutter, Russell. "Poetry, Imagination, and Technical
Writing." College English 47 (1985): 698-712.
Explains how literature teachers are more prepared
to teach technical writing than they believe they are.

105. Stratton, Charles R. "Technical Writing: What It Is
and What It Isn't." Journal of Technical Writing and
Communication 9 (1979): 9-16.
States that both teachers of literature and teachers
of composition are technical writers because they do
"job-related writing" (10). Discusses similarities and
differences between teaching technical writing and
teaching either literature or composition.

106. Walter, John A. "Teaching Technical Writing: Resources and Strategies." Technical Writing Teacher 8
(1980): 29-33.
Offers suggestions for how teachers of literature
can prepare to teach technical writing. These suggestions include self-help; professional societies; institutes,
seminars, and workshops; and internships and graduate
courses. Briefly discusses strategies for the teaching
of technical writing.

1.5 Content of the Technical Writing Course

107. Ad Hoc Committee of the American Business Communica-
tion Association. "Student Evaluation of the Basic
Course in Business Communication." Journal of Business
Communication 12.4 (1975): 17-24.
Describes a survey to determine students' attitudes
toward basic courses in business communication.

108. Aldrich, Pearl G. "Adult Writers: Some Factors That
Interfere with Effective Writing." Research in the

Teaching of English 16.3 (1982):   298-300.
Describes a survey conducted to determine writing
problems of adults.

109.   Allred, Hilda F., and Joseph F. Clark.   "Written Communication: Problems and Priorities."   The Journal of
Business Communication 15.2 (1978):   31-36.
Describes a survey conducted to determine what
business professionals and business teachers consider
problems in writing done by college graduates.   Suggests changes in curriculum based on the findings of
the survey.

110.   Anderson, Edward.   "On-the-Job Writing for Vocational
Students: Research and Implications."   Teaching English in the Two-Year College 9 (1983):   231-35.
Discusses studies conducted to determine on-the-job
writing requirements of vocational students.

111.   Anderson, Paul V.   "Career Opportunities for Teachers
of Technical Writing: A Survey of Programs in Technical Communication."   Journal of Technical Writing and
Communication 8 (1978):   175-92.
Presents the results of a mail survey of twenty-nine
schools that offered programs in technical communication to determine what career opportunities exist in the
field of technical communication.

112.   _____.   "What Technical and Scientific Communicators Do: A Comprehensive Model for Developing Academic Programs."   IEEE--Transactions on Professional
Communications PC-27 3 (1984):   161-67.
Presents a model that helps educators determine what
technical and scientific communicators need to know.

113.   Andrews, J. Douglas, and Norman B. Sigband.   "How
Effectively Does the 'New' Accountant Communicate?
Perceptions by Practitioners and Academics."   The
Journal of Business Communication 21.2 (1984):   15-24.
Describes a survey to examine employers' thoughts
concerning the communicating abilities of recent accounting graduates.   Offers suggestions for changes
in accounting curriculums to improve the communication
skills of accounting students.

114. Arnoff, Craig E. "Business Communication: Funda-
      mentals and the Future." <u>ABCA Bulletin</u> 46.2 (1983):
      39-41.
           Examines changes that are influencing the area of
      business communication. Evaluates the present state
      of business communication and suggests actions that
      should be taken to improve education in business com-
      munication.

115. Bataille, Robert R. "Writing in the World of Work:
      What Our Graduates Report." <u>College Composition and
      Communication</u> 33 (1982): 276-80.
           Describes a survey conducted to determine how much
      writing was done on the job by recent college graduates.

116. Bates, Gary D. "Upgrading Written Communication--
      Your Firm's and Your Own." <u>IEEE--Transactions on
      Professional Communications</u> PC-27 2 (1984): 89-92.
           Discusses that state of written communication in
      business and suggests ways written communication can
      be improved.

117. Bowes, J. E., and K. R. Stamm. "Science Writing
      Techniques and Methods." <u>Journal of Environmental
      Education</u> 10.3 (1979): 25-28.
           Describes a review of literature to determine views
      concerning science writing.

118. Brockman, R. John. "Advisory Boards in Technical
      Communications' Programs and Classes." <u>Technical
      Writing Teacher</u> 9 (1982): 137-46.
           Describes a survey conducted to determine what
      role advisory boards have in designing and implement-
      ing technical communication programs and classes.

119. Burkett, Allan R., and Susan B. Dunkle. "Technical
      Writing in the Undergraduate Curriculum." <u>Journal of
      Chemical Education</u> 60.6 (1983): 469-70.
           Discuss problems that chemistry and engineering
      majors face in the technical writing classroom.

120. Carlisle, E. Fred. "Teaching Scientific Writing Human-
      istically: From Theory to Action." <u>English Journal</u>
      67.4 (1978): 35-39.
           Argues that science and English are united by the

concern with interpersonal communication. Offers a
theoretical basis and concludes with a pedagogical approach emphasizing this link.

121. Carter, Ron. "Technical Writing: A framework for
      Change." Technical Writing Teacher 7 (1979): 39-42.
          Argues that technical writing should be taught as a
      "process" not as a series of formulas for writing various
      types of papers. States that students need to know
      how "to design the most effective possible response to
      any communication situation" (39).

122. Casari, Laura E. "Required: Three Hours in Techni-
      cal Communication--Paradigm for a Paradox." IEEE--
      Transactions on Professional Communications PC-27 3
      (1984): 116-120.
          Examines the problems of industry's demand for
      engineers with good communication skills and academia's
      failure to provide engineering students with adequate
      training in communication.

123. Coney, Mary B., and others. "Technical Writing in
      the English Department: An Outside Perspective."
      ADE Bulletin 79 (1984): 30-42.
          Discusses ways to improve technical writing instruc-
      tion in English departments.

124. Cronin, Frank C. "Writing Requirements in a Large
      Construction Firm." Technical Writing Teacher 10
      (1983): 81-85.
          Reports results of "taped interviews with fifteen up-
      per and middle management executives of a very large
      construction firm located in the Northeast" (81). Focuses
      on "hiring procedures, the writing requirements of
      entry level employees and upper management, and ad-
      vice for college students preparing to enter the business-
      industrial world" (81).

125. Dobrin, David N. "What's the Purpose of Teaching
      Technical Communication?" Technical Writing Teacher
      12 (1985): 146-60.
          Examines the purpose of teaching Technical Communi-
      cation and advances both moral and pragmatic goals.
      Suggests curricular and administrative changes to stress
      these moral goals and "develop a unity of purpose" (159).

126. Dodge, Kristen A. "The Writing Morass: Redefining the Problem." The Journal of Business Communication 17.4 (1980): 19-32.
Discusses the problem of poor writing in terms of differing definitions of poor writing as well as deciding who should teach writing and how.

127. Dunwoody, Sharon. "From a Journalist's Perspective: Putting Content into Mass Media Science Writing." English Journal 67.4 (1978): 44-47.
Discusses "social/political/economic" (45) considerations of science writers. Explains the implications of such considerations on the teaching of science writers.

128. Ellman, Neil. "Science in the English Classroom: Teaching Ideas." English Journal 67.4 (1978): 63-65.
Argues for an interdisciplinary connection between science and language. Offers specific suggestions for how this connection can be made.

129. Feinberg, Susan, and Jerry Goldman. "Content for a Course in Technical Communication." Technical Communication 32.2 (1985): 21-25.
Describes a study to determine what those in industry, government and academia considered important for designing a course in technical communication.

130. Feinberg, Susan, and Jerry I. Goldman. "Technical Writing Attitude Measurement and Instructional Goals." IEEE--Transactions on Professional Communications PC-27 3 (1984): 155-60.
Describes an instrument for measuring students' attitudes toward technical writing. Presents findings from the use of this instrument.

131. Fleischhauer, F. W. "Standardizing Writing in Business and Industry." Journal of Technical Writing and Communication 6 (1976): 275-92.
Argues for standardizing writing in business and industry and suggests methods for doing so.

132. _____. "Meeting the Writing Needs of Business and Industry." Journal of Technical Writing and Communication 11 (1981): 223-232.
States that technical writing courses should teach

students how to write to meet the day-to-day activities
within an organization.

133. Ford, Arthur L.   "Technical Writing and the Liberal
      Arts School." Journal of Technical Writing and Com-
      munication 9 (1979):   271-79.
      Examines issues concerning the teaching of technical
      writing in liberal arts colleges.

134. Galle, William P., Jr.   "Don't Teach Employment Skills
      in the Business Communications Course." ABCA Bul-
      letin 46.3 (1983):   3-6.
      Explains why teaching students how to obtain jobs
      should not be a part of the business communication
      course.

135. Goldstein, Jone Rymer.   "Trends in Teaching Technical
      Writing." Technical Communication 31.4 (1984):   25-34.
      Discusses changes that are taking place as to how
      technical writing is viewed and taught.

136. Golen, Steven; Celeste Powers; and M. Agnes Titke-
      meyer.   "How to Teach Ethics in a Basic Business Com-
      munication Class--Committee Report of the 1983 Teach-
      ing Methodology and Concepts Committee, Subcommittee
      I." Journal of Business Communication 22.1 (1985):
      75-83.
      The authors define ethics as what is "right" and
      "good" according to the cultural situation. They offer
      guidelines to "assist" one in making ethical decisions.
      States that basic business communication classes should
      help students learn how to make ethical decisions. De-
      scribes a teaching unit, including objectives, activities,
      and sources, that instructors can use to introduce the
      issue of ethics in the basic business communications
      classroom.

137. Gould, Jay R.   "Evaluation of a Master's Program in
      Technical Communication--Results of a Questionnaire."
      Journal of Technical Writing and Communication 7 (1977):
      55-73.
      Describes a survey of graduates from Rensselaer's
      Master's Program in Technical Communications to find
      out how well prepared they were to handle communi-
      cation situations on the job.

138. Gwiasda, Karl E. "Of Classrooms and Contexts: Teaching Engineers to Write Wrong." IEEE--Transactions on Professional Communications PC-27 3 (1984): 149-51.
  States that writing teachers and engineering teachers should work together to teach engineering students how to be better writers.

139. Halpern, Jeanne W. "What Should We Be Teaching Students in Business Writing?" The Journal of Business Communication 18.3 (1981): 39-53.
  Describes six "categories" of content that should be offered in a business writing course.

140. Harris, Elizabeth. "In Defense of the Liberal Arts Approach to Technical Writing." College English 44 (1982): 628-32.
  Argues that writing for business and technology should not be severed from the liberal arts.

141. Hart, Maxine B. "Integrating Office Technology and Procedures into Business Communication's Courses." ABCA Bulletin 46.4 (1983): 34-37.
  Examines the role that developing technology has in business communication courses. Describes some courses designed to incorporate issues concerning new technology into the business communication classroom.

142. Hart, Sara. "Key Philosophical Factors in Developing a Business Communication Curriculum." The Journal of Business Communication 13.4 (1976): 47-57.
  Describes a study of college level institutes to assist in the development of a curriculum for a business communication course.

143. Harwood, John T. "Freshman English Ten Years After: Writing in the World." College Composition and Communication 33 (1982): 231-83.
  Describes a survey conducted to determine the writing skills needed on the job by college graduates of Christopher Newport College in Virginia.

144. Haselkorm, Mark P. "A Pragmatic Approach to Technical Writing." Technical Writing Teacher 11 (1984): 122-24.
  Argues that "Technical Writing should be limited to

the study and teaching of pragmatic conventions relevant to actual technical communication situations" (122).
Uses proposal writing as an example of how this limita-
tion allows "objective criteria" to be used to help in
determining "the limits of technical writing" (124).

145. _____. "Linguistic Boundaries of Technical Writing."
Technical Writing Teacher 11 (1983): 26-30.
Argues that the boundaries of technical writing
should be defined so that the technical class differs
from other writing classes and so researchers can better
study the field of technical writing. By using concepts from formal linguistics, the article concludes that
the field of technical writing is concerned with "the
study of Pragmatic Structure and Rhetoric for those
communication situations that are actually encountered
by people working in technical contexts" (29).

146. Henson, Leigh. "Identifying Effective Writing Exercises for Lower-Division Technical Writing Courses."
Journal of Technical Writing and Communication 14
(1984): 307-16.
Describes a survey conducted to determine "the
instructional aims of lower-division technical writing"
and what assignments can be used in conjunction with
the aims (308).

147. Holloway, Watson L. "Looking for Mr. Jones: The
Ideal of the American Artifex." Technical Writing
Teacher 12 (1985): 23-27.
Examines changing opinions toward technical students
and discusses how these changing opinions affect teachers
of technical students.

148. Hunt, Peter, and Sarah Wilkinson. "Technical Communication: The Academic Dilemma." Journal of Technical
Writing and Communication 15 (1985): 35-42.
Examines the factors that hinder the improvement of
training in technical communication that the academic
community offers.

149. Journet, Debra. "Rhetoric and Sociobiology." Journal
of Technical Writing and Communication 14 (1984): 339-
50.
Reexamines the traditional view that science is

"objective and non-rhetorical." Shows that science is
rhetorical by discussing scientific articles that present
scientific debate concerning sociobiology.

150. Kowal, John Paul. "Training Tomorrow's Writers."
Journal of Technical Writing and Communication 5 (1975):
181-85.
States that the role of effective communication in-
creases as America's economy becomes a "service" rather
than a "production" economy. Examines the effect that
jargon has on communication. Suggests that as part of
their training writers must see examples of good writing,
understand how these documents were developed and
apply this understanding to their own writings which
teacher will evaluate.

151. La Roche, Mary G. "Technical Writing as an Alternative Rhetoric for Composition Courses." Technical
Writing Teacher 8 (1981): 73-76.
States that technical writing is "essentially a rhetoric, a form of writing, rather than material or subject matter for writing" (73). Shows how learning to
present technical subjects helped students writing
classes produce better writing. Suggests that both "bel-
letristic rhetoric" and "technical rhetoric" can be used
together in freshman English classes.

152. Lewis, Phillip V. "The Status of 'Organizational Communication' in Colleges of Business." The Journal of
Business Communication 12.4 (1975): 25-28.
Describes a survey conducted to determine if students were beginning to take more courses in organizational communication than traditional writing courses.

153. Mac Intosh, Fred H. "Where Do We Go from Here?"
Journal of Technical Writing and Communication 8 (1978):
139-45.
Suggests ways to improve both the presentation of
and teaching of technical writing.

154. Marling, William. "'The Vague Aches of Interns.'"
College English 45 (1983): 690-94.
Examines "the conflict between the interns' expectations of the real world and the kind of writing that the
real world needs" (690) by discussing an internship

program that the author developed. Finds that creativity does not have to be stifled by the routine,
"formuaic" writing demanded in the real world.

155. McCarron, William E. "Changing the Technical Writing
      Paradigm." Journal of Technical Writing Communication
      15 (1985): 27-33.
          Proposes changes in the traditional view concerning
      how technical writing should be taught.

156. Meis, Ben H. "Cybernetica Qualified: Technical vs.
      Creative Writing" Community and Junior College Journal
      54.6 (1984): 22-24.
          Compares kinds of writing taught in basic composi-
      tion courses to that taught in technical writing courses.
      Also compares creative writers to technical writers.

157. Mitchell, Ruth. "Shared Responsibility: Teaching
      Technical Writing in the University." College English
      43 (1981): 543-55.
          Examines the battle between English teacher and
      engineering educators over who should teach technical
      writing. Discusses attempts made to define technical
      writing to devise a new "taxonomy" of technical writing
      that assists in curriculum design by guiding "the degree
      of cooperation between teachers of the discipline and
      teachers of writing" (552).

158. Pearsall, Thomas E. "Current University Programs in
      Technical Communication." Technical Communication
      22.1 (1975): 16-18.
          Discusses a three-day conference that focuses on
      current university programs in technical communication.

159. Raymond, Michael W. "Giving English the Business
      Cliches and Misconceptions." CEA Forum 11.3 (1981):
      12-14.
          Suggests that educators rethink the purpose of
      teaching English and consider traditional views in order
      to consider the role of English instruction in relation
      to the business world.

160. Rivers, William E. "The Place of Business Writing in
      English Departments: A Justification." ADE Bulletin
      65 (1980): 27-31.

Examines the changes made in the teaching of business writing courses.

161. Ross, Frederick C., and Michell H. Jarosz. "Integrating Science Writing: A Biology Instructor and an English Teacher Get Together." English Journal 67.4 (1978): 51-55.

The authors describe how they, from different backgrounds, became interested in the other's field. Then, they explain how this joint interest led them to develop writing courses that included the joint efforts of science and English teachers.

162. Ryan, Michael, and Sharon L. Dunwoody. "Academic and Professional Training Patterns of Science Writers." Journalism Quarterly 52.2 (1975): 239-46, 290.

Examines changes in the training of science writers. Finds that more science writers are starting to pursue academic degrees.

163. Schindler, George E., Jr. "Why Engineers and Scientists Write as They Do--Twelve Characteristics of Their Prose." IEEE--Transactions on Professional Communications PC-18 (1975): 5-10.

Discusses some of the major writing problems of engineers and scientists. Explains what causes these problems.

164. Schneider, Carolyn. "Debriefing Academically Trained Writers." ABCA Bulletin 44.4 (1980); 4-5.

Argues that much of what is taught in college writing courses is not relevant in the business world.

165. Sides, Charles H. "Technical Writing: Implications for Compositional Skills Development." Technical Writing Teacher 8 (1981): 54-57.

Argues that technical writing should not be taught as a craft course but as a writing course that involves the use of rhetoric. States that technical writing is discourse and discusses technical writing as "a method of writing development," "a heuristic for perceiving and organizing experience," and as "a pragmatic alternative to the traditional ... approaches to composition" (55).

166.   _____. "What Should We Do with Technical Writing?"
        Engineering Education 70.7 (1980):   743-44.
            Argues that technical writing should be taught in
        English departments.

167.   Smeltzer, Larry R.   "An Analysis of Communication
        Course Content for MBA Students."   The ABCA Bulletin
        47.3 (1984):   28-33.
            Describes a study conducted to determine what con-
        tent should be included in MBA communication courses.

168.   Snipes, Wilson Currin.   "Language, Composition, and
        Literature in Technical Education."   Teaching English
        in the Two-Year College 6.1 (1979):   13-19.
            Suggests that technical writing courses and general
        writing courses should have the same goal: the develop-
        ment of thoughtful, communicating adults.

169.   Souther, James W.   "What's New in Technical Writing."
        Technical Writing Teacher 8 (1980):   34-38.
            Discusses the "challenges and satisfactions" (35)
        that are in store for technical writing teachers.   Also
        lists the concepts that should be included in a technical
        writing course.

170.   Sparrow, W. Keats.   "Syllabus Revision Through Co-
        operative Education: Adapting Courses to the 'Real
        World.'"   Journal of Cooperative Education 18.1 (1981):
        94-98.
            Explains how working in conjunction with co-op em-
        ployers to design course content makes for more rele-
        vant business communication courses.

171.   Steigler, C. B.   "Trends in the Content of the Basic
        Business Communication Course: An Editor's Viewpoint."
        ABCA Bulletin 40.2 (1977):   14-19.
            Evaluates the content of basic business communication
        courses taught at the college level.

172.   Stevenson, Dwight W.   "Mapping the Unexplored Area:
        Developing New Courses and Coherent Programs in
        Technical Communication."   Journal of Technical Writing
        and Communication 8 (1978):   193-206.
            Presents a procedure that teachers of technical writ-
        ing and administrators can use to determine "needs" for
        and "objectives" of technical writing courses.

173. Stine, Donna, and Donald Skarzenski. "Priorities for
the Business Communication Classroom: A Survey of
Business and Academe." The Journal of Business Communication 16.3 (1979): 15-30.
Discusses a survey conducted to determine what
those in business and academia consider to be important
communication skills and how these skills should be
taught in business courses.

174. Storms, C. Gilbert. "Programs in Technical Communication." Technical Communication 31.4 (1984): 13-20.
Provides an analysis of a survey of the literature
concerning programs in technical communication. Offers recommendations for improving existing programs.

175. _____. "What Business School Graduates Say About
the Writing They Do at Work: Implications for the
Business Communication Course." ABCA Bulletin 46.4
(1983): 13-18.
Describes a survey of graduates from Miami University's School of Business Administration to determine
what writing tasks they face on the job. Suggests
ways to improve business communication courses.

176. Sullivan, Jeremiah J. "The Importance of a Philosophical 'Mix' in Teaching Business Communication." The
Journal of Business Communication 15.4 (1978): 29-37.
Examines various philosophies of education and their
role in the teaching of business communication.

177. Swenson, Dan H. "Relative Importance of Business
Communication Skills for the Next Ten Years." The
Journal of Business Communication 17.2 (1980): 41-49.
Describes a study to determine how important com-
munication skills will be for students graduating within
the next ten-year period.

178. Whitburn, Merrill. "Technical Communication: An Unexplored Area for English." ADE Bulletin 45 (1975):
11-14.
Argues that English departments must teach business
and technical communication to meet the needs of future
students.

179. Wilkes, John. "Science Writing: Who? What? How?"

English Journal 67.4 (1978): 56-60.
 Shows how English teachers, with little or no background in science, can teach science students to write
by having them write to non-scientists about scientific
topics. Suggests using dialogue writing to begin with
then expanding these dialogues into expository or argu-
mentative papers.

180. Wilkinson, A. M. "Organizational Writing vs. Business
      Writing." The ABCA Bulletin 47.3 (1984): 37-39.
       Discusses the difference between organizational writ-
      ing and business writing. Outlines a course in organ-
      izational writing.

181. Wyld, Lionel D. "Practica and Internships for Tech-
      nical Writing Students." Technical Writing Teacher 10
      (1983): 86-90.
       Examines employment opportunities available to
      students through "graduate internships," "intersession
      programs and practica," "co-op programs," "workshops
      and other employment." Discusses how technical writing
      teachers can help place students in the working world.

1.6 Preparations for Teaching Technical Writing

182. Barnum, Carol M. "English Professors as Technical
      Writers: Experience Is the Best Teacher." ADE Bul-
      letin 76 (1983): 32-34.
       Describes the experiences of an English teacher who
      became a technical writer in industry. Discusses how
      such an experience could benefit all technical writing
      teachers.

183. Day, William R. "A 'Business Internship' for Technical
      Writing Teachers." Teaching English in the Two-Year
      College 4 (1977): 43-45.
       Suggests that teachers of technical writing should
      obtain experience in the business world in order to be
      more effective teachers.

184. 184. Dobrin, David N. "What's Difficult About Teach-
      ing Technical Writing." College English 44 (1982):
      135-40.
       Describes some of the difficulties that teachers of

technical writing face. Focuses particularly on the difficulties arising from the fact that in many cases "the
teacher of technical writing is teaching the student to
perform for his or her peers in a particular technical
community, a community of which the teacher is not
himself a member" (137). Offers suggestions for over-
coming these difficulties.

185. Harris, John S. "Learned Any Science or Technology
      Lately?" Technical Writing Teacher 9 (1982): 188-95.
          Quizzes teachers of technical writing to determine
      what they have learned from teaching technical writing
      students.

186. Hartley, James. "Eighty Ways of Improving Instruc-
      tional Text." IEEE--Transactions on Professional Com-
      munications PC-24 (1981): 17-28.
          Details a list of guidelines for writing better instruc-
      tional material.

187. Hull, Keith N. "Notes from the Besieged, or Why Eng-
      lish Teachers Should Teach Technical Writing." College
      English 41 (1980): 876-83.
          Argues that English teachers should teach technical
      writing. Explains how English teachers can be pre-
      pared for teaching technical writing.

188. Kellner, R. S. "A Question of Competency." ABCA
      Bulletin 45.3 (1982): 5-10.
          Discusses the qualifications of teachers of business
      and technical writing.

189. Knapper, Arno F. "Good Writing--A Shared Responsi-
      bility." The Journal of Business Communication 15.2
      (1978): 23-27.
          Argues that faculty members from all disciplines
      should work together to teach communication skills to
      students.

190. McCarron, William E. "Opinion: In the Business World
      and in Academe: The English Teacher in the 1980's."
      College English 41 (1980): 812-21.
          Examines the changing role of English teachers.
      Discusses developing technological and literary issues
      that English teachers need to consider.

191. Muller, John A.   "What Consultation and Freelance
      Writing Can Do For You and For Your Students."   Con-
      ference on CCC, Kansas City, Missouri, 31 March-2
      April 1977.
          Suggests that technical writing teachers should seek
      experience in the world outside academia in order to
      better teach their students.

192. Nelson, Kate.   "Tilting at Windbags: My Losing Battle
      with Business Jargon."   English Journal 73.5 (1984):
      61-62.
          Describes an English teacher's experience as a tech-
      nical writer in industry.

193. Pickett, Nell Ann.   "Preparing to Teach a Technical
      Writing Course in the Two-Year College."   Technical
      Writing Teacher 7 (1979):   2-7.
          Explains how teachers new to the teaching of tech-
      nical writing can obtain materials and make "contacts
      with local people."   Includes lists of possible textbooks;
      textbook publishers, professional associations; confer-
      ences, and journals.

194. Reinbold, R.   "Teach Technical Writing? How? A Re-
      Tooling Model for Beginning Technical Writing Teachers."
      Pennsylvania Council of Teachers of English Conven-
      tion, Champion, Pennsylvania, 14-15 October 1977.
          Discusses seven strategies for preparing to teach
      technical writing.

195. Rothmel, Steven Zachary.   "Opportunities for the Tech-
      nical Writing Teacher."   Technical Writing Teacher 6
      (1979):   106-107.
          Briefly describes some of the opportunities available
      to technical writing teachers.   States that the "technical
      writing teacher is an integral part of the successful
      melding" of science, business, and the humanities (107).

196. Stalnaker, Bonny J.   "But I've Never Taught Scientific
      and Technical Writing Before!"   English Journal 67.4
      (1978):   48-50.
          Describes how someone new to the teaching of scien-
      tific/technical writing can survive.   Explains the skills
      that students will need to develop in the course and
      what knowledge the teacher must have to teach the

course. Then, offers a way that teachers can plan a
course to meet the students' needs.

197. "The Technical Writing Controversy." The ABCA Bul-
      letin 46.2 (1983):   4-11.
          Presents responses made by various authors to R.
      S. Kellner's "The Degenerative of Technical Writing:
      A Question of Teacher Competency" that appeared in
      the September 1982 edition of the ABCA Bulletin.

198. Walter, John A. "ATTW: In Retrospect and in Pros-
      pect." Technical Writing Teacher 7 (1980):   91-93.
          Examines what the ATTW has accomplished and sug-
      gests what the ATTW can do in the future.

199. Walter, Marie-Louise. "Think Now--Write Later: The
      Triumphs and Traumas of a New Teacher of Technical
      Writing." Journal of Technical Writing and Communica-
      tion 9 (1979):   147-152.
          Describes the author's experience as a new teacher
      of technical writing. Discusses the rewards that come
      from teaching technical writing and suggests some teach-
      ing methods and tools that worked for the author.

2.1 Community and Two-Year Programs

200. Carter, Ron. "Technical Writing in the Two-Year College." Teaching English in the Two-Year College 4
(1978): 222-23.
  Examines the state of technical writing in two-year
colleges.

201. Hemphil, P. D. "Teaching Business Communications in
Community College." The Journal of Business Communication 13.1 (1975): 41-49.
  Describes who the community college student is and
what is most important for this student to learn about
business communication. Lists the material that should
be covered in the business communication classroom and
suggests methods for best presenting the materials.

202. Lehman, Anita J. "Technical Writing: English for a
Lay Audience." Teaching English in the Two-Year College 6 (1979): 67-70.
  Describes the "design, implementation, and evaluation" of a course entitled "Writing for Industry." Suggests that 1) students can learn from a course "based
on clearly defined levels of competence," 2) "attitudes
toward the importance of communication can be made
more positive," and 3) "writing is a discipline that can
be learned" (p. 69).

203. Pickett, Nell Ann. "Preparing to Teach a Technical
Writing Course in the Two-Year College." Conference
on CCC, Minneapolis, Minnesota, 5-7 April 1979.
  Presents suggestions for how one can prepare to
teach a technical writing course at a two-year college.

204. Silver, Marilyn B. "Business Communication: A Com-

munity College Approach. ABCA Bulletin 44.3 (1981): 21-26.

Discusses the characteristics of community college students. Describes a curriculum used at Delaware Technical and Community College. Explains the advantages of using a curriculum that brings "job-related and interdisciplinary materials into the composition classroom" (p. 22).

205. Skelton, Terry. "Research As Preparation: Determining the Instructional Needs of the Two-Year Student." Conference on CCC, Denver. 30 March-1 April 1978.

Describes a study to determine the on-the-job writing needs of graduates from two-year colleges.

2.2 Graduate Programs

206. Feinberg, Susan, and Irene Pritzker. "An MBA Communications Course Designed by Business Executives." The Journal of Business Communication 22 (1985): 75-83.

Describes a model for a graduate-level business communications course. Explains how the model is based on suggestions made by business executives.

207. Journet, Debra. "Preparing Teachers of Technical Writing." Technical Writing Teachers 11 (1983): 1-6.

Describes a graduate seminar on the teaching of technical writing. The article discusses how theory and pedagogy concerning technical and scientific communication were covered in the seminar. The article also provides a syllabus for the 14-week seminar.

208. Masse, Roger E. "Editing in Technical Communication: Theory and Practice in Editing Processes at the Graduate Level." Annual Meeting of CCCC, Detroit, Michigan, 17-19 March 1983.

Describes a graduate-level technical-editing course offered at New Mexico State University.

209. Munter, Mary. "Trends in Management Communication at Graduate Schools." The Journal of Business Communication 20.1 (1983): 5-11.

Discusses changes that are being made in graduate-level business communication courses.

210. Smeltzer, Larry. "An Integrated Approach for a Grad-
     uate Course in Business Communication." ABCA Bul-
     letin 44.3 (1981): 27-29.
         Describes "a graduate seminar that integrates written,
     oral, and organizational communication in one course"
     (p. 27). Discusses the needs of master-level and
     doctoral-level students that take the seminar and ex-
     plains how this seminar meets the students' needs.

211. Wharton, Raena, and Merrill Whitburn. "Texas A&M
     University's Course for Training Technical Writing
     Teachers." Technical Writing Teacher 8 (1980): 27-28.
         Recounts the development and current status of
     Texas A&M's graduate course on the teaching of tech-
     nical communication. Discusses the curriculum of the
     course and research projects available to students in
     the course.

2.3 In-House Training Programs

212. Baim, Joseph. "In-House Training in Report Writing:
     A Collaborative Approach." ABCA Bulletin 40.4 (1977):
     5-8.
         Explains what is necessary to develop a program for
     teaching writing within a business organization.

213. Daley, Paul W., and Herman A. Estrin. "Teaching An
     In-House Public Speaking Course." Journal of Tech-
     nical Writing and Communication 12 (1982): 271-77.
         Outlines an in-house course in public speaking.
     Includes a syllubus and suggested assignments.

214. David, Carol, and Donna Stine. "Measuring Skill Gain
     and Attitudes of Adult Writers in Short Courses."
     ABCA Bulletin 47.1 (1984): 14-20.
         Describes a study to determine how much adult
     writers improve their writing skills through short courses.

215. Denton, L. W. "In-House Training in Written Com-
     munication: A Status Report." The Journal of Business
     Communication 16.3 (1979): 3-14.
         Describes a study to determine cost, content, and
     effectiveness of in-house training in written communica-
     tion.

216. Dickson, Donald R. "Planning for the In-House Writing Seminar." ABCA Bulletin 45.2 (1982): 14-16.
Explains what needs to be done to prepare a six- to eight-week writing program to be taught in-house.

217. Fleishchhauer, F. W. "Meeting the Writing Needs of Business and Industry." ABCA Bulletin 45.5 (1982): 36-41.
Presents a business and technical writing course taught in business and industry.

218. Hennelly, John. "Business Writing Where It Counts." English Journal 73.2 (1984): 85-88.
Describes an in-house writing workshop for supervisors in industry.

219. Hertz, Vivienne. "Traveling and Teaching--A New Delivery System for Technical Writing." Technical Writing Teacher 5 (1978): 88-90.
Describes a technical writing course taught at an air force base.

220. Khachaturian, Armen, and Herman A. Estrin. "Teaching In-House Technical Communication Courses." ABCA Bulletin 46.2 (1983): 42-48.
Outlines an in-house workshop designed to teach effective writing and oral presentation skills. Includes syllabus, assignments, and course evaluation sheet.

221. Rothwell, Williams J. "Developing an In-House Training Curriculum in Written Communication." The Journal of Business Communication 20.2 (1983): 31-45.
Explains why a series of in-house writing courses is more effective than a one-time writing workshop. Presents a sample curriculum for a four-course workshop.

222. Wallisch, Bill. "Using Television Technology to Teach Technical Writing." International Technical Communication Conference, Los Angeles, May 1979.
Describes the teaching of technical writing at the U.S. Air Force Academy.

223. Weber, Max. "An Advanced In-House Course in Technical Writing." Technical Writing Teacher 3 (1976):

92-107.
   Discusses the teaching of an advanced technical
writing course offered at the Argonne National Laboratory. Suggest textbooks, describes assignments, and
promotes peer critiquing.

224.   Whalen, Tim. "Developing an In-House Business and
       Technical Writing Course." IEEE--Transactions on
       Professional Communications PC-26 (1983):   160-61.
          Discusses main considerations for anyone interested
       in developing an in-house course.

225.   _____. "Techniques for Developing an Effective In-
       House Course in Business and Technical Writing." ABCA
       Bulletin 42.3 (1980):   27-28.
          Presents a "plan" for developing an in-house business
       and technical writing course.

2.4 Interdisciplinary Programs

226.   Callazzo, Louis J., III. "Scientific Writing Course."
       Technical Writing Teacher 2 (1975):   14-16.
          Describes a three-credit course entitled "Scientific
       Writing" which is offered at the Philadelphia College of
       Pharmacy and Science. Discusses the major topics
       covered and the assignments given in the class.

227.   Carlisle, E. Fred, and Jack B. Kinsinger. "Scientific
       Writing: A Humanistic and Scientific Course for Science
       Undergraduates." Journal of Chemical Education 54
       (1977):   632-34.
          Describes a writing course developed by the English
       and science departments at Michigan State University.

228.   Dick, John A. R., and Robert M. Esch. "Dialogues
       Among Disciplines: A Plan for Faculty Discussions of
       Writing Across the Curriculum." College Composition
       and Communication 36 (1985):   178-192.
          Discusses ways that faculty members from different
       disciplines can work together to plan writing across
       the curriculum programs.

229.   Echman, Martha. "An Interdisciplinary Program in
       Technical Communications: Problems Encountered"

Technical Writing Teacher 6 (1979): 87-90.
Examines three types of problems that arise when colleges and universities try "to establish a program in technical communications" (p. 87):
"1. Those involving society's need for better technical communication;
"2. Those involving the business community specifically, and
"3. Those involving the academic community specifically" (p. 87).

230. Faigley, Lester, and Kristine Hansen. "Learning to Write in the Social Sciences." College Composition and Communication 36 (1985): 140-49.
Discusses the nature of writings done in the social sciences. Suggests what teachers must do to meet the writing needs of students from many disciplines.

231. Houston, Linda S. "An Integrated Approach to Technical Writing." Technical Writing Teacher 8 (1981): 71-72.
Describes how a technical writing course is set up at Ohio State University Agricultural Technical Institute to allow students, writing teachers and technical faculty to work together.

232. Potvin, Janet H., and Robert L. Woods. "Teaching Technical Communication at the Graduate Level: An Interdisciplinary Approach." Journal of Technical Writing and Communication 13 (1983): 235-246.
Describes an interdisciplinary approach used to teach technical communication at the University of Texas at Arlington.

233. Schall, Maryan. "Profile of the Technical Communication Student at the University of Minnesota." Technical Communication 23.3 (1976): 2-6.
Describes an interdisciplinary course in Metallurgical Engineering and English taught at Ohio State University.

234. Schenck, Eleanor M. " 'How Can We Help Technical Writing Students?' Reading Centers Ask." Journal of Technical Writing and Communication 6 (1976): 123-28.
Describes a survey of technical writing instructors conducted to determine the instructors' views concerning

the value of reading and their experiences with reading
programs. Suggests ways that technical writing teachers
and reading instructors can work together to teach
students communication skills.

235.   Silver, Marilyn B.   "Technical Report Writing in the
       Community College: An Interdiciplinary Approach."
       Technical Writing Teacher 9 (1982):   173-78.
           Describes an interdisciplinary technical writing
       course for a community college.   Provides guidelines
       for setting up such a program.

236.   Thilsted, Wanda H.   "An Interdisciplinary Report Writ-
       ing Course."   Technical Writing Teacher 2 (1975):   1-3.
           Describes how an interdisciplinary report writing
       course was set up by the English Department at Okla-
       homa State University.   Details the content of the
       course.   Explains how this course "opened up some
       lines of communication between departments on campus"
       (p. 3).

237.   Van Pelt, William V.   "Teaching Technical Writing for
       the Computer Sciences."   Technical Writing Teacher 10
       (1983):   189-94.
           Describes a course offered at the University of
       California, Santa Cruz which helps to prepare students
       for writing computer documentation.   Explains how such
       a course can prepare students by introducing students
       to the "specialized context" that they will be placed in
       when writing computer documentation and by teaching
       students how to cope with "the rapid advance of com-
       puter technology" (p. 189).

238.   Vik, Gretchen N.   "What We Need to Teach in a Basic
       Writing Skills Course."   ABCA Bulletin 44.1 (1981):
       25-28.
           Describes "an adjunct course between the Study
       Skills Department at the college of Business" (p. 25)
       taught at San Diego State University.   Discusses which
       errors in grammar and punctuation give business stu-
       dents the most problems.

239.   Weightman, Frank C.   "Renewing Professional Credi-
       bility: Technical Writing in English Departments."
       Technical Writing Teacher 10 (1983):   182-85.

Describes a "broadly-based program in language study and writing" (p. 182) that exists at Memphis State University. List both undergraduate and graduate courses offered, explains how students can earn "Writing and Publication Specialist Certificate," and discusses "on-site training workshops offered to businesses."

2.5 Undergraduate Programs

240. Alexander, Clara. "Teaching Technical and Business Writing: Strategies and Evaluation." Teaching English in the Two-Year College 12 (1985): 113-17.
      Details the contents of a course which prepares students for the oral and written communication assignments they will face on the job.

241. Allred, Hilda F., and Joseph F. Clark. "Written Communication Problems and Priorities." The Journal of Business Communication 15.2 (1978): 31-35.
      Describes a study to determine what recent college graduates thought about their preparedness for handling communication demanded on the job.

242. Alred, Gerald J. "Developing a Technical Writing Theory and Pedagogy Course in an English Department." Journal of Technical Writing and Communications 10 (1980): 337-45.
      Describes a course on the teaching of business and technical writing. Includes a sample syllabus.

243. Ballantyne, Ver Don W. "Syllabus for English 215--Exposition and Report Writing, Brigham Young University." Technical Writing Teacher 3 (1976): 60-83.
      Provides a detailed syllabus of a writing class offered at Brigham Young University. Explains the philosophy behind the choice of course content and details the objectives and assignments for each unit of the course.

244. Bostian, Frieda F. "Technical Writing--'Very Useful Stuff.'" IEEE--Transactions on Professional Communications PC-27 (1984): 121-25.
      Describes a technical writing course taught at Virginia Polytechnic Institute and State University.

245. Brockmann, R. John. "Taking a Second Look at Tech-
      nical Communications Pedagogy." Journal of Technical
      Writing and Communications 10 (1980): 283-91.
         Describes problem areas that exist in undergraduate
      technical communications curricula and explains how
      these problems developed. Suggest that the problem
      can be solved by emphasizing revision and "cooperative
      aspects of composition," increasing students' confidence
      in producing graphics, and encouraging the use of oral
      communication among small groups.

246. Broome, Michael. "Technical Writing Is a B.I.G.
      Course." College Composition and Communication 27
      (1976): 55-57.
         Describes how the author approached the teaching
      of a technical writing course at Columbia College, South
      Carolina.

247. Brostoff, Anita. "Individual Instruction: Tailored to
      Particular Needs and Specific Disciplines." Conference
      On CCC, Denver, 30 March-1 April 1978.
         Describes the communication skills center at Carnegie
      Mellon University.

248. Broughton, Bradford B. "A Key Course to Unlock
      Communication: Letter Writing." IEEE--Transactions
      on Professional Communications PC-24 (1984): 193-
      96.
         Describes a letter writing course taught at Clarkson
      University in Potsdam, New York.

249. Brown, Michael R. "Writing and Science: A Freshman
      Writing Course for Science Majors." Conference on
      CCC, St. Louis, 13-15 March 1975.
         Describes a course offered at Western Michigan Uni-
      versity.

250. Corbett, Edward P. J. "A Collegiate Writing Program
      for the 1980's." ADE Bulletin 78 (1984): 20-23.
         Describes a writing program that addresses tech-
      nological advances that affect students in the 1980's.

251. Cornelius, Fred. "Taking the 'Bull' By the Horns:
      Teaching the Fundamentals of Business Writing." Teach-
      ing English in the Two-Year College 5 (1978): 47-56.

Discusses the development of a curriculum for a business writing course that considers 1) what students have been taught in other writing courses and 2) the principles of writing in conjunction with the forms of business writing and expectations of business executives. Finally, the author suggests assignments that are effective in the business writing course.

252. Cramer, Carmen. "Go Out and Prosper, Technocrats: Technical Writers and Rhetorical Translations." Conference of CCC, Minneapolis, Minnesota, 21-23 March 1985.
Describes the technical writing major offered at the University of Southwestern Louisiana.

253. Daniel, Carter A. "Communication Course First, Remedial English Afterward." ABCA Bulletin 46.4 (1983): 42-43.
Suggests the advantages of offering a remedial English course for those students who struggled to master basic writing skills in the business communication course.

254. De Beaugrande, Robert. "Audience and Focus in Technical Writing." Report prepared at Ohio State University, 1978.
Describes a training program for technical writers.

255. Dorrell, Jean, and Betty Johnson. "A Comparative Analysis of Topics Covered in Twenty College-Level Communication Textbooks." ABCA Bulletin 45.3 (1982): 11-16.
Evaluates 20 college-level business communication textbooks.

256. Dowell, Earl E.; David Schuelke; and Chew Wah Chung. "Evaluation of A Bachelor's Program in Technical Communication: Results of a Questionnaire." Journal of Technical Writing and Communication 10 (1980): 195-200.
Presents the result of a questionnaire designed to evaluate the undergraduate technical communication program in the Department of Rhetoric at the University of Minnesota, St. Paul.

257. Dudley, Juanita Williams. "A New String for an Old

Bow--Business Writing with Journalistic Applications."
ABCA Bulletin 40.3 (1977):   7-10.
Describes a business communication course offered
at Purdue University. Explains the importance of having students read and produce business publications.

258. Dunwoody, Sharon, and Ellen Wartella. "A Survey of
the Structure of Science Writing Courses." Association
for Education in Journalism Convention. Seattle, Washington, 13-16 August 1978.
Describes a survey of science writing courses to
determine the content offered by such courses.

259. Easom, Roger D. "Lining Up the Georges Without Shoot-
ing Them Down: Composition for Technical Students."
Teaching English in the Two-Year College 7 (1981):
143-46.
Suggests that the teaching of practical, real-world
writing begins in the Freshman composition classroom.
This allows students to combine writing and their technical fields. Students learn by seeing the need for
writing and are not "shot down" by their deficiencies
in writing that are often the focus of traditional freshman composition classrooms.

260. Farkas, David. "A Course and Curriculum in Advanced
Technical Communication." Conference on CCC. San
Francisco, California, 18-20 March 1982.
Describes a technical communication course offered
at West Virginia University.

261. Foster, Gretchen. "Technical Writing and Science
Writing: Is There a Difference and What Does It Mean?"
Conference on CCC. New York, New York, 29-31
March 1984.
Describes an experimental course designed to help
students understand the nature of technical writing.

262. Glassman, Myron, and Ann E. Farley. "AACSB Accredited Schools' Approach to Business Communication
Courses." The Journal of Business Communication 16.3
(1979):   41-48.
Describes the results of a survey to determine the
content of business communication courses offered by
schools accredited by the American Assembly of Collegiate Schools of Business.

263.   Gloe, Esther M. "Setting Up Internships in Technical
       Writing." Journal of Technical Writing and Communica-
       tion 13 (1983): 7-27.
            Presents issues that should be addressed when
       establishing internship programs.

264.   Hackos, Joann T. "Teaching Problem--Solving Strate-
       gies in the Technical Communication Classroom." IEEE--
       Transactions on Professional Communications PC-27
       (1984): 180-84.
            Describes a writing curriculum designed at the
       Colorado School of Mines to teach students to be better
       problem solvers in order to become better technical
       communicators.

265.   Hemphill, P. D. "Teaching Business Communications
       in Community College." The Journal of Business Com-
       munication 13.1 (1975):   41-49.
            Describes the characteristics of communication college
       students. Presents techniques for teaching business
       communication to these students.

266.   Henson, Leigh. "A Node on the Status of Lower-division
       Technical Writing." Technical Writing Teacher 11 (1984):
       89-93.
            Discusses the results of surveying 740 individual
       members of the Association of Teachers of Technical
       Writing to determine what institutions offer courses in
       lower-division technical writing and what instructors
       consider to be the aim of lower-division technical writ-
       ing.

267.   Howard, Lionel A., Jr. "A Survey of Technical Com-
       munication Programs in U.S. Colleges and Universities."
       IEEE--Transactions on Professional Communications
       PC-26 (1983):   113-16.
            Examines various technical communication programs
       offered in the United States.

268.   Howard, Lionel A., Jr. "A Survey of Technical Com-
       munication Programs in United States Colleges and Uni-
       versities--1984." IEEE--Transactions on Professional
       Communications PC-27 (1984):   172-176.
            Presents a list of technical communication programs.
       Categorizes the programs according to their emphases.

269. Hull, Leon C. "Internships in Technical Writing: A
      Sponsor's View." Technical Communication 24.1 (1977):
      7-9.
         Addresses complaints of students' and sponsors'
      concerning internships.

270. Johnson, Laurence J. "Courses in Writing Beyond the
      University: A Caveat." Journal of Technical Writing and
      Communication 7 (1977): 105-20.
         Compares curricula of college-level and in-house writ-
      ing courses. Suggests ways to improve writing curricula
      to improve writing skills among professionals.

271. Kellner, Robert Scott. "The Lexicon of Science: A
      Course in Technical Terminology." Journal of Techni-
      cal Writing and Communication 15 (1985): 55-61.
         Proposes that a course in technical terminology be
      offered to students trying to master technical communica-
      tion. Presents the design for such a course.

272. Kelton, Robert W. "The Role of a Private Research
      Foundation in a Technical Writing Program." Journal
      of Technical Writing and Communication 10 (1980): 59-64.
         Describes how interaction between Battelle Memorial
      Institute and Ohio State University allows for better
      training of technical writers.

273. Lacombe, Joan M., and Joanne G. Kane. " 'Write On':
      Teaching Effective Written Communication." Business
      Education World 57.3 (1977): 24-25.
         Describes how business communication courses are
      taught at Bay Path Junior College in Longmeadow, Mass.,
      so that students find the material relevant.

274. Lamphear, Lynn. "Business Writing at the University of
      Missouri, St. Louis." ABCA Bulletin 43.3 (1980): 20-21.
         Describes the content of a one-semester business
      writing course offered by the English Department at
      the University of Missouri, St. Louis.

275. Loeb, Helen M. "Answering Industry's Needs: An
      Internship Training Program for Technical Writers."
      Annual Meeting of International Technical Communication
      Conference. Boston, Massachusetts, 5-8 May 1982.
         Describes an internship program offered at North-
      eastern University in Boston.

276.   Matthiesen, Eric E., and Millicent L. Pogge. "A Com-
       parative Analysis of Undergraduate Programs in Or-
       ganizational and Business Communication." The Journal
       of Business Communication 16.4 (1979):   31-43.
           Describes a survey conducted to determine what
       college-level programs are available in business com-
       munication and to compare the contents of such pro-
       grams.

277.   Matulich, Loretta. "Contract Learning in the Traditional
       Technical Writing Class." Annual Convention of Ameri-
       can Association of Community and Junior Colleges. New
       Orleans, Louisiana, 24-27 April 1983.
           Describes the use of contract learning in a technical
       writing class at Clackamas Communication College in
       New York.

278.   Mitchell, John H. "Three Prescriptive Approaches to
       Teaching Factual Communications." Technical Writing
       Teacher 4 (1976):   7-12.
           Discusses "three graduate and undergraduate pilot
       courses" that were offered by the English department
       of the University of Massachusetts in Amherst. States
       that the prescriptive nature of the courses appealed
       to both faculty, "trained in literature," and students
       who had to take the courses.

279.   Murphy, Karl M. "The Basic Technical and Business
       Writing Course at Georgia Tech." ABCA Bulletin 40.3
       (1977):   5-7.
           Describes a business writing course offered at
       Georgia Tech. Discusses assignments used in the
       course.

280.   Nelson, Ronald J. "Planning the Adventure: An Ap-
       proach to the Technical Writing Course at a Technical
       College." Technical Writing Teacher 8 (1981):   66-70.
           Offers a four-step approach for planning a "tech-
       nical writing course at a technical college" so that the
       course is "a fulfilling experience for both student and
       teacher" (p. 66).

281.   Ober, Scot, and Alan P. Wunsh. "The Status of Busi-
       ness Communication Instruction in Postsecondary Insti-
       tutions in the United States." The Journal of Business

Communication 20.2 (1983):   5-16.
    Describes a survey of 344 postsecondary institutions
"to determine the status of business communication instruction in postsecondary institutions throughout the
United States" (p. 5).   Shows that the majority of institutions have business communication courses within
the school of business administration.

282.   Pearsall, Thomas E.   "Beyond the Basic Course: Build-
       ing a Program in Technical Communication."   Technical
       Writing Teacher 8 (1980):   39-42.
           Explains why a need exists for more programs in
       technical communication.   Describes types of programs
       and offers curriculum guidelines for the programs.

283.   _____.   "Building a Technical Communication Program."
       Conference on CCC.   Dallas, Texas, 26-28 March 1981.
           Describes the steps involved in planning and setting
       up a program in technical communication.

284.   Ramsden, Patricia A.   "Utilizing Workshops in Technical
       Composition."   Available through ERIC Clearinghouse.
       Published January 1980.
           Describes a course in technical writing taught at
       Madisonville Community College in Kentucky.   Shows
       the advantages of combining lecture and workshop
       methods of teaching.

285.   Ruehr, Ruthann.   "Some Characteristics and Writing
       Problems of Technically Oriented Students."   Conference
       on CCC.   New York, New York, 29-31 March 1984.
           Describes writing problems of technical students at
       Michigan Technological University.

286.   Sawyer, Thomas M.   "A Syllabus for a Course in Scien-
       tific and Technical Communication."   Technical Writing
       Teacher 2 (1975):   17-23.
           Provides a detailed syllabus designed for and de-
       scribes assignments required in a course in Scientific
       and Technical Communication taught in the College of
       Engineering at Michigan.

287.   Seigle, Natalie R.   "Syllabus: Oral and Written Business
       Communications."   ABCA Bulletin 42.2 (1979):   15-17.
           Presents a syllabus for a business communication
       course that includes both oral and written assignments.

288. Silver, Marilyn B. "Teaching Writing to Adult Learners: Using Job-Related Materials." Improving College and University Teaching 30.1 (1982): 33-37.
   Discusses the teaching of communication skills to adults at Delaware Technical and Community College.

289. Skelton, Terrance. "Determining the Instructional Needs of the Two-Year Student." Teaching English in the Two-Year College 7 (1980): 57-64.
   Examines the needs of vocational students. Provides suggestions for designing a technical communication course that will meet these needs.

290. Skelton, Terrance, and Deborah C. Andrews. "An Advanced Course in Business and Technical Publications." Journal of Technical Writing and Communication 15 (1985): 215-25.
   Describes a course offered at the University of Delaware to teach strategies for managing document production, designing and editing numerous types of documents, working in teams, handling projects in conjunction with businesses and the university, and mastering computer skills.

291. Southard, Sherry G. "The Internship Program at Oklahoma State University." Annual Meeting of the Council for Programs in Technical and Scientific Communication. Lincoln, Nebraska, 7-8 April 1983.
   Describes the internship program offered at Oklahoma State University.

292. Soutter, Florence. "Analysis of Business Communication Curricula." ABCA Bulletin 41.2 (1978): 24-25.
   Describes a survey conducted to determine what should be taught in the business communication classroom. Finds that basic writing principles should be emphasized more.

293. Stoddard, Ted D. "Business Communication as a Competency-Based General Education Course." The Journal of Business Communication 17.5 (1980): 50-60.
   Examines the merits of general education. Describes a general education program that exists at Brigham Young University. Discusses how business communication fits into the general education program and presents

the "outcomes" of "teaching business communication as
a competency-based general education course" (p. 57).

294.   Tebeaux, Elizabeth.   "Redesigning Professional Writing
       Courses to Meet the Communication Needs of Writers in
       Business and Industry."   College Composition and Com-
       munication 36 (1985):   419-28.
              Discusses the types of professional writing courses
       usually offered at colleges and universities.   Examines
       the needs of writers on the job and suggests changes
       in the basic writing course in order to meet the writers'
       needs.

295.   Van Arsdel, Rosemary T.   "The University of Puget
       Sound's Writing Institute."   ADE Bulletin 76 (1983):
       35-37.
              Discusses a writing program that trains English
       majors for careers as technical writers.

296.   Whitburn, Merrill D.   "New Program Development at
       Texas A&M."   Journal of Technical Writing and Com-
       munication 7 (1977):   347-48.
              Discusses changes being made in the English depart-
       ment at Texas A&M University to better prepare stu-
       dents in the field of technical communication.

297.   Wilkinson, C. W.   "Grammar, Theory, and Principles
       in the Basic College Course in Business Communication
       (What? Whether? How Much?)"   ABCA Bulletin 40.3
       (1977):   1-4.
              Offers guidelines for teaching a course in business
       writing.

298.   Winkler, Victoria M., and Jeanne L. Mizuno.   "Advanced
       Courses in Technical Writing: Review of the Literature
       (1977-84)."   Technical Writing Teacher 12 (1985):   33-
       49.
              Reviews literature available on "Advanced Courses
       in Academia" and "Advanced Courses in Business and
       Industry."   Concludes that technical writing teachers
       "concentrate on technical communication as opposed to
       focusing narrowly on writing" (p. 46).

3.1 Stressing the Importance of Communication Skills

299. Allen, J. W., Jr. "Introducing Invention to Technical Students." Technical Writing Teacher 5 (1978): 45-49.
     Discusses how to introduce students to invention techniques. Examines the usefulness of invention techniques for technical writing students.

300. Brown, Ann E. "A Practicum for the Technical Writing Classroom." Technical Writing Teacher 12 (1985): 102-06.
     Describes an exercise developed by the Writing Assistance Program of the School of Engineering at North Carolina State University. The exercise involves comparing original and rewritten versions of a "real-world" memo. Finds that students respond favorably to such "practical demonstrations."

301. Cheshire, Ardner. "Teaching Invention: Using Topical Categories in the Technical Writing Class." Technical Writing Teacher 8 (1980): 17-21.
     Explains how to use the "topics" from D'Angelo's A Conceptual Theory of Rhetoric to teach invention to technical writing students.

302. Cook, John M. "The Technical Writing Student as Editor." Technical Writing Teacher 10 (1983): 114-17.
     Suggests that technical writing teachers need to help students understand that the roles of the technical writer and technical editor are "compatible and continuous." Explains that "in order for writers to understand why they are revising they must understand what an editor does and for what reasons" (p. 114).

303. Ewald, Helen Rothschild, and Donna Stine. "Speech

Act Theory and Business Communication Conventions."
The Journal of Business Communication 20.3 (1983):
13-25.
    Relates H. P. Grice's "maxims" of conversation to
written business communication. Explains how principles of speech act theory can be applied when assessing
failed business communication attempts.

304. Farkas, David K. "An Invention Heuristic for Business
and Technical Communication." ABCA Bulletin 44.4
(1981): 16-19.
    Describes "a heuristic designed specifically for busi-
ness and technical professionals" (p. 16). Explains
how the heuristic will help writers meet the goals of
professional communication. Also explains how instructors can teach this heuristic to students.

305. Feinberg, Susan. "Visual Patterns: An Experiment
with Technically Oriented Writers." Technical Communication 31.1 (1984): 20-22.
    Describes an experiment which used visuals to teach
technically oriented writers how to better organize information.

306. Garver, Eugene. "Teaching Writing and Teaching Virtue." The Journal of Business Communication 22.1
(1985): 51-73.
    Argues "that virtue can be taught" (p. 62). States
that by teaching students how to use "practical reasoning" (p. 63) in their writing and by teaching students
"the subject-matter of practical reasoning" (p. 64),
teachers help students learn to incorporate virtue in
their writings.

307. Georgopoulos, Chris J., and Voula C. Georgopoulos.
"From University Term Papers to Industry Technical
Reports--An Attempt to Bridge the Existing Gap."
IEEE--Transactions on Professional Communications
PC-27 (1984): 144-148.
    Suggests ways to better prepare students for com-
munication tasks they will face on the job.

308. Gibbs, Meada, and others. "How to Teach Effective
Listening Skills in a Basic Business Communication
Class." ABCA Bulletin 48.2 (1985): 30-34.

Presents a teaching unit designed to improve students' listening skills. Gives an outline for the unit and describes suggested classroom activities.

309. Hall, Dennis R. "The Role of Invention in Technical Writing." Technical Writing Teacher 4 (1976): 13-24.
Argues that technical writing students should understand "invention as a generative system" (p. 14). States that use of invention helps students view technical writing "as a process which promotes the extension of science and technology" instead of "a form into which science and technology contract" (p. 18).

310. Hamilton, D. "Writing Science II." Journal of Education 162 (1980): 96-113.
Describes the author's examination of examples of science writing in order to determine how teachers can better train science writers.

311. Hart, Maxine Barton. "Communication: Teaching Business Writing with Fewer Writing Assignments." Business Education Forum 34.2 (1979): 24-25, 27.
Describes a study conducted at Baylor University Hankamer School of Business to find new methods of teaching writing that would satisfy students and meet course objectives. Finds that the inclusion of communication theory and the reduction of writing assignments produced students who wrote as well as students taught by a traditional approach.

312. Hays, Robert. "Quirk Topics Enliven Technical Writing Classes." Journal of Technical Writing and Communication 14 (1984): 43-48.
Explains the concept of "quirk topics" and shows how these topics can be used to produce interesting technical reports. Lists twenty possible "quirk topics."

313. Iyasere, Marla Mudar. "Teaching Technical Writing: Coping with Students Misconceptions and Evaluation Anxieties." Journal of Technical Writing and Communication 15 (1985): 259-66.
Examines views that students have toward technical writing. Presents suggestions for showing students the importance of possessing good writing skills and for developing grading methods that are relevant for the students.

314. Jack, Judith. "Teaching Analytical Editing." Tech-
      nical Communication 31.1 (1984): 9-11.
           Discusses a strategy for teaching students how to
      edit effectively.

315. Jackson, Deborah. "Communication: Provide Solid
      Foundation for Good Business Communications." Busi-
      ness Education Forum 32.3 (1977): 31, 33.
           Explains that business communication can only be
      taught if attention is given to reading, writing, listen-
      ing, and speaking. States that a communication course
      should help students improve their vocabulary and
      develop more effective written and oral communication
      skills. Suggests teaching methods that will help achieve
      these goals.

316. Johnston, Jean. "Don't Leave Your Language Alone!"
      Business Education Forum 31.2 (1976): 9-11.
           Discusses the events that led to a decline in the
      English language. Offers suggestions as to how edu-
      cators can stop this decline and teach students to write
      and speak correctly.

317. Jones, Daniel R. "A Rhetorical Approach for Teaching
      the Literature of Scientific and Technical Writing."
      Technical Writing Teacher 12 (1985): 115-25.
           Describes how the rhetorical approach to scientific
      and technical writing can be applied to the classroom.
      Lists numerous texts that can be used when teaching
      rhetorical topics and provides examples of how these
      texts can help generate discussions.

318. Little, C. B. "Teaching by Examples." Teaching
      Sociology 9 (1982): 401-06.
           Presents a method for teaching students how to im-
      prove their writing by showing them examples of good
      writing.

319. Lunsford, A. A. "Classical Rhetoric and Technical
      Writing." College Composition and Communication 27
      (1976): 289-91.
           Suggests a method for using classical rhetoric in
      the technical writing course to show students the rele-
      vance of technical writing.

320. Lynn, Steven Walton. "Kinneavy, Mathes, Mumford, and Lynn: Teaching the Classificatory Mode in Technical Writing." Journal of Technical Writing and Communication 10 (1980): 115-24.
    Uses Kinneavy's definition of the classificatory mode to describe how Mathes' "contextual editing" can be used in the teaching of technical writing. Describes how to explain the concepts of classification and contextual editing to students and how to have students apply these concepts to Mumford's "Machines, Utilities, and 'The Machine'" and to their own writings.

321. Matthews, Anne L., and Patricia G. Moody. "Communication: Communication Skills for Career-Minded Students." Business Education Forum 32.7 (1978): 19-20.
    Explains that all teachers in business departments are responsible for teaching their students to have better reading, writing, speaking, and listening skills. Offers suggestions as to how teachers can help students develop these skills.

322. McGalliard, Roy A. "Teaching Writing Is a Relevant Act." Technical Writing Teacher 4 (1977): 102-04.
    Examines the teaching of writing and determines what constitutes successful teaching.

323. Miller, Barbara. "The Campus Publication as an Instructional Tool." Technical Writing Teacher 3 (1976): 129-34.
    Suggests that students can learn editorial skills by serving as "editors of a campus publication" (p. 129). Describes which skills students learn and explains how working through the editorial process helps them master these skills.

324. Pearce, Frank. "Desirable Writing Skills for Personnel in Business and Industry." Teaching English in the Two-Year College 3 (1976): 29-31.
    States that the best way to teach writing is to focus on basic skills.

325. Radloff, David M. "Replacing Some Assumptions in the Teaching of Technical Writing." Journal of Technical Writing and Communication 9 (1979): 141-45.
    States that the term paper is seldom applicable to

the real world. Argues that "idea grids" are more
practical and useful than outlines. Explains that ideas
are created in a "linear, sequential, and measured flow"
and shows how an idea grid can be used to match the
more natural flow of ideas. Finds that technical writing texts are not helpful and suggests alternatives to
using the texts.

326. Roundy, Nancy. "A Program for Revision in Business
      and Technical Writing." The Journal of Business Com-
      munication 20.1 (1983):   55-66.
            Describes a program for revision used by the author
      in her business and technical writing course. Explains
      how the program helps students improve their written
      documents.

327. Roundy, Nancy, and Charlotte Thralls.   "Modeling the
      Communication Context: A Procedure for Revision and
      Evaluation in Business Writing." The Journal of Busi-
      ness Communication 20.3 (1983):   27-46.
            Describes a four-part model of the communication
      context. Discusses the relationship between "situation,
      audience, purposes and uses, and explanation of writ-
      ing decisions" (p. 28). Applies the model to revision
      and evaluation.

328. Simon, Judith C. "Practice What You Preach--A Posi-
      tive Approach in Teaching Business Communication."
      ABCA Bulletin 42.3 (1979):   31-32.
            States that teachers should use a positive approach
      in the classroom to help students understand the im-
      portance of communicating information positively.

329. Sims, Barbara B. "Increasing Student Involvement in
      Technical Writing Classes." Technical Writing Teacher
      2 (1975):   9-11.
            Discusses the importance of student involvement in
      the classroom. Suggests assignments that attempt to
      increase student involvement in technical writing classes.

330. Smeltzer, Larry R., and Kittie W. Watson.   "A Test of
      Instructional Strategies for Listening Improvement in a
      Simulated Business Setting." The Journal of Business
      Communication 22.4 (1985):   33-42.
            Describes a study that compares four methods for

improving listening skills. Finds that use of methods for improving listening skills did improve students' performances.

331. Sparrow, W. Keats. "Motivating Recalcitrant Business and Technical Writing Students." Journal of Technical Writing and Communication 5 (1975): 305-09.
Describes an approach for convincing technical writing students that knowing how to write well is important.

332. Suchan, James. "Managing the Business Communications Classroom's Organizational Environment." ABCA Bulletin 47.4 (1984): 13-15.
Defines characteristics of the business communication classroom environment. Suggests ways for creating an environment that facilitates learning.

333. Sullivan, Patricia A. "Teaching the Writing Process in Scientific and Technical Writing Classes." Technical Writing Teacher 8 (1980): 10-16.
Argues that until textbooks provide discussion of the writing process teachers should stress that process. Provides a "system" for incorporating discussions of the writing process into discussions of various types of documents. Gives examples and applications of this "system."

334. Titen, Jennifer. "Application of Rudolf Arnheim's Visual Thinking to the Teaching of Technical Writing." Technical Writing Teacher 7 (1980): 113-18.
Explains what Rudolf Arnheim's theory of visual thinking has to offer the technical writing classroom.

335. VanDeWeghe, Richard. "Writing Models, Versatile Writers." The Journal of Business Communication 20.1 (1983): 13-23.
Presents five models of the composing processes that can be used to teach students in business and technical writing classrooms.

336. Varner, Iris I. "Communicating Report Evaluations to Students." ABCA Bulletin 44.1 (1981): 21-22.
Explains how discussing writing assignments in terms of "the mechanics of language, the logical use of the

language, and the analysis of the business situation"
(p. 21) helps students understand evaluations of their
writings.

337. Vaughn, Jeannette W. "The Basics of English: Foun-
      dation for Success in Technical Communication." Tech-
      nical Writing Teacher 12 (1985): 19-22.
         Examines the problem of poor English skills from the
      perspectives of technical communication experts, execu-
      tives, students, and instructors. Finds that all groups
      recognize the importance of knowing the fundamentals
      of English. Provides suggested techniques that educa-
      tors can use to help students master the fundamentals
      of English.

338. Vielhaber, Mary Elizabeth. "Coping with Communica-
      tion Anxiety: Strategies to Reduce Writing Apprehen-
      sion." ABCA Bulletin 46.1 (1983): 22-24.
         Discusses the causes of writing apprehension. Sug-
      gests ways that teachers and students can eliminate
      some of the factors that produce writing apprehension.

339. Zoerner, C. E., Jr. "Teaching the Vanquished to
      Write." The Journal of Business Communication 13.1
      (1975): 33-39.
         Suggests ways for encouraging beginning business
      communication students so that they continue to be
      motivated throughout the business communication course.

3.2 Simulating the "Real World" Environment

340. Boris, Edna Z. "The Interview in a Business Writing
      Course." ABCA Bulletin 41 (1978): 23-24.
         Explains how to teach interviewing with the context
      of the job hunt in college-level technical writing courses.
      Provides suggested assignments.

341. Loris, Michelle Carbone. "A Collaborative Learning
      Model: The Rhetorical Situation as a Basis for Teach-
      ing Business Communication." ABCA Bulletin 46.1
      (1983): 25-27.
         Describes a teaching unit which has groups of stu-
      dents assume the roles of "fictional" companies. Pro-
      vides sample assignments for the unit.

342. McCoy, Joan, and Harlan Roedel. "Drama Is the Class-
      room: Putting Life in Technical Writing." Technical
      Writing Teacher 12 (1985): 11-17.
          Explains how to devise and use skits to dramatize
      "business situations in which memos, letters and re-
      ports are written" (p. 11). Believes that these simula-
      tions "give students a better sense of what it means to
      write for real readers for business reasons" (p. 11).
      Provides sample skits and sample student writing.

343. McLeod, Alan. "Stimulating Writing Through Job Aware-
      ness." English Journal 67.8 (1978): 42-43.
          Explains how teachers can "stimulate" student writ-
      ing by giving students assignments that will involve
      them with the business community. Provides a list of
      some possible assignments.

344. Rutter, Russell. "Through a Different Looking-Glass:
      Technical Writing to Train the Imagination." Journal
      of Technical Writing and Communication 11 (1981): 121-
      29.
          Argues that technical writing courses should "simu-
      late 'real life' so that students view the course as a
      "job situation" (p. 122). Uses a proposal assignment
      as an example to show how the teacher can establish
      an environment where he/she serves as supervisor
      and students function as employees.

345. Speck, Bruce. "A Simulated Approach to Business
      Writing." ABCA Bulletin 46.4 (1983): 38-41.
          Describes a simulated business environment that the
      author developed in order to provide students with
      relevant writing assignments.

346. Tebeaux, Elizabeth. "Achieving the Essential Goals of
      Technical Writing--Role Playing Won't Do It!" ABCA
      Bulletin 41.3 (1978): 1-3.
          Discusses the effectiveness of role-playing. Reminds
      teachers to concentrate on students' writing problems.

3.3 Using the Case Method

347. Butler, Marilyn S. "A Reassessment of the Case Ap-
      proach: Reinforcing Artifice in Business Writing

Courses." <u>ABCA Bulletin</u> 48.3 (1985): 4-7.
Argues against the use of the case approach to
teach students in business and technical writing. Suggests instead that students be allowed to draw from
their own experiences.

348. Cochran, Daniel S., and Kendrick Gibson. "Putting a
Square Peg into a Round Hole: Communication Models
and Their Application." <u>The Journal of Business Com-
munication</u> 17.1 (1979): 27-36.
Explains how teachers can combine an integrative
communication model, system theory and the case method
to produce higher levels of learning in the communication classroom.

349. Couture, Barbara, and Jone Goldstein. "How to Develop and Write a Case for Technical Writing." Conference on CCC. Dallas, Texas, 26-28 March 1981.
Describes how to develop a case to be used in the
teaching of technical writing.

350. Hays, Robert. "Case Problems Improve Technical Writing Courses and Seminars." <u>The Journal of Technical
Writing and Communication</u> 6 (1976): 293-98.
Discusses the advantages of using case problems as
instructional tools. Provides guidelines for preparing
cases. Includes a sample case.

351. Kingsley, Lawrence. "The Case Method as a Form of
Communication." <u>The Journal of Business Communica-
tion</u> 19.2 (1982):   39-50.
Argues that the case method is ineffective unless
the cases are well written. Describes the problems that
arise when cases are not written well. Finally, describes what constitutes a well written case.

352. Mascolini, Marcia, and Caryl P. Freeman. "The Case
for Cases." <u>Technical Writing Teacher</u> 7 (1980): 125-
26.
Examines the advantages of using cases in technical
writing classrooms. Provides sample cases.

353. McCleary, William J. "A Case Approach for Teaching
Academic Writing." <u>College Composition and Communica-
tion</u> 36 (1985): 203-12.

Describes a case approach to teaching and explains how to write effective case assignments.

354. Robbins, Jan C. "Training the Professional Communicator: The Case Study Method." The Journal of Business Communication 12.3 (1975): 37-45.
Suggests using the case study method to help students become expert communicators.

355. Roundy, Nancy. "Heuristics for Invention in a Technical Writing Class." Technical Writing Teacher 6 (1983): 200-09.
Discusses two types of invention procedures that technicians use on the job to generate content and to select content. Explains how technical writing teachers can have students "mirror" these invention techniques by asking students to "evolve cases for every document they write" (p. 202).

356. Welford, Chester L. "Teaching Audience in Technical Writing or the Technical Writing Teacher as Weekend Novelist." Technical Writing Teacher 7 (1979): 12-14.
Explains how case studies allow for the teaching of audience analysis in conjunction with teaching other skills. Discusses a sample case.

3.4 Using Journal Assignments

357. Bratchell, D. F., and G. A. Mitchell. "The Assessment of Communication Skills." Journal of Technical Writing and Communication 5 (1975): 39-46.
Examines the importance of adequately training technical teachers. Evaluates various methods of assessing student writing and suggests the use of journal writing as an effective way to assess student writing.

358. Eisenberg, Anne. "Laboratory Notebooks: Current Teaching Applications." Journal of Technical Writing and Communication 12 (1982): 213-17.
Explains how the keeping of laboratory notebooks can be turned into effective technical writing assignments.

359. Flatley, Marie. "Improve Business Writing Skills with

Reactive Writing."   ABCA Bulletin 47 (1984):   25-26.
Explains how students can improve their writing
skills by keeping a journal in which they compile written
responses to articles they read.

360.   Goldstein, Jone Rymer, and Elizabeth L. Malone.   "Journals on Interpersonal and Group Communication: Fa-
cilitating Technical Project Groups."   Journal of Technical Writing and Communication 14 (1984):   113-31.
Explains how students' oral and written skills can
be improved by requiring students to keep journals on
their group activities.

3.5 Using Group Projects

361.   Covington, David H.   "Making Team Projects Work in
Technical Communication Courses."   Technical Writing
Teacher 11 (1984):   100-04.
Describes the author's use of team projects in his
technical communication class.   Finds that team projects
are effective "means of stimulating professional communication activity" (p. 104).   Offers some guidelines,
based on his experience, for the use of team projects
in the classroom.

362.   Hagaman, John A.   "Practice What You Preach: Techniques for Teaching Personalized Business Letters to
Established Business Writers."   ABCA Bulletin 41.2
(1978):   17-19.
Discusses the effectiveness of using peer critiques
in a business letter writing workshop.

363.   Hughes, Robert S., Jr.   "Developing Business Communications Skills with the Overhead Projector: Student-
Centered Techniques that Work."   ABCA Bulletin 47.2
(1984):   21-24.
Explains how students can use overhead projectors
to facilitate group editing sessions.

364.   Meyers, G. Douglas.   "Adapting Zoellner's 'Talk-Write'
to the Business Writing Classroom."   ABCA Bulletin
48.2 (1985):   14-16.
Describes a teaching method which involves pairing
students so that the students talk with each other about

assigned writing tasks, take notes as they talk, and incorporate these notes into drafts of their papers.

365. Sills, Caryl Klein. "Adapting Freewriting Techniques and Writing Support Groups for Business Communication." ABCA Bulletin 48.2 (1985): 12-14.
    Explains how Peter Elbow's theories concerning peer response and freewriting can be applied to the business communication classroom.

366. Todd, Mavis M. "Communication: Learning to Write for the Reader." Business Education Forum 34.8 (1980): 17-18.
    Describes a teaching method for helping students become more reader-oriented. Explains how students work with partners to write and evaluate letters of invitation.

3.6 Teaching Style

367. Allison, Desmond. "Forms, Meanings, and Uses in Language Teaching: Some Problems and Possibilities." English Language Teaching Journal 37 (1983): 154-57.
    Discusses approaches for teaching grammar in scientific writing.

368. Dobrin, David N. "Teaching Infinitive Phrases." Technical Communication 32.3 (1985): 44-45.
    Discusses the author's method for teaching students how to correctly use infinitive phrases.

369. Douglas, George H. "The Informational Sketch--A Major Vice of the Technical Writer." Journal of Technical Writing and Communication 12 (1982): 7-13.
    States that although brief, concise writing is a virtue of technical writing, this kind of writing can become too "compact," producing sentences that are hard to follow and understand. Suggests methods that teachers can use to correct "compactness" that produces writing that has no continuity.

370. Estrin, Herman. "The Classics of Scientific Literature in the Technical Writing Course." Teaching English in the Two-Year College 12 (1985): 303-06.

Discusses the use of Walter Miller's "What Can the Technical Writer of the Past Teach the Technical Writer of
Today?" to teach the concept of technical style.

371.   Hamed, Charles. "Use Numbers to Demonstrate the
       Control of Emphasis in Written Messages." ABCA Bul-
       letin 46.1 (1983):   13-15.
       Explains how to use a list of ten numbers to teach
       students the importance of using "positioning, spacing,
       repeating, and flagging" to emphasize messages in busi-
       ness writing.

372.   Kent, Thomas L. "Paragraph Production and the
       Given-New Contract." The Journal of Business Com-
       munication 21.4 (1984):   45-66.
       Discusses the theory of the "Given-New Contract"
       and presents pedagogical applications that can be de-
       rived from this theory.

373.   _____. "Six Suggestions for Teaching Paragraph
       Cohesion." Journal of Technical Writing and Communi-
       cation 13 (1983):   269-74.
       Draws from information theory and linguistics to
       present suggestions for teaching paragraph cohesion.

374.   Miller, Carolyn R. "Rules, Context, and Technical
       Communication." Journal of Technical Writing and
       Communication 10 (1980):   149-58.
       Takes "rules theory" from anthropology and linguis-
       tics and applies it to technical communication. Shows
       how rules theory can be used to help communicators
       better understand contexts and thus produce more ef-
       fective messages within various contexts.

375.   Minor, Dennis E. "A Concept of Language for Techni-
       cal Writing Students." Journal of Technical Writing
       and Communication 8 (1978):   43-51.
       Describes an approach for teaching students about
       the use of language in technical writing.

376.   Olson, Gary A. "Sentence Combining." Technical
       Communication 31.1 (1984):   p. 38.
       Discusses the use of sentence combining as a teach-
       ing tool.

377. Orth, Melvin F. "Color Their Prose Gray." IEEE--
      Transactions on Professional Communications PC-18
      (1975): 64-66.
          Suggests ways for improving the writing of scientists
      and engineers.

378. Pearce, C. Glenn. "Communication: Help Students
      Produce More Understandable Writing." Business Edu-
      cation Forum 32.4 (1978): 27, 29.
          Offers suggestions for how teachers can help stu-
      dents improve their writing. Encourages teachers to
      stress the importance of topic sentences, introductions
      and conclusions, headings, and illustrations.

379. Roberts, David D. "Teaching Generalization and Sup-
      port." Technical Communication 31.2 (1984): 28.
          Describes a method for teaching students how to
      support generalizations in their technical writing.

380. Rosner, Mary, and Terri Paul. "Using Sentence Com-
      bining in Technical Writing Classes." Technical Writing
      Teacher 10 (1982): 35-40.
          Provides sentence combining exercises that can be
      used in the technical writing classroom.

381. Roundy, Nancy. "Teaching Arrangement in Business
      Messages." ABCA Bulletin 46.4 (1983): 21-26.
          Details a "comprehensive program" (p. 22) that can
      be used to teach arrangement of information within a
      text.

382. Stagrin, Michael. "Sentence-Combining, Conceptual
      Sophistication, and Precision in Technical Exposition."
      Technical Writing Teacher 7 (1979): 28-34.
          Explains how sentence-combining exercises can en-
      hance instruction given in technical writing classrooms.

383. Steffey, Marda Nicholson. "'Paragraph Building' as a
      Teaching Tool." ABCA Bulletin 46.2 (1983): 23-26.
          Describes a method for teaching students how to
      construct effective paragraphs. Includes sample exer-
      cises.

384. Tebeaux, Elizabeth. "Using the Extended Definition
      Paper to Teach Organization." Journal of Technical

Writing and Communication 10 (1980): 3-10.
Explains how teaching the extended definition helps
reinforce discussion of audience analysis and shows
students various says to organize information.

385. Tesch, Robert C., Sr. "An Analysis of Three Instructional Strategies Used in Teaching Business Communication Grammar and Usage." The Journal of Business
Communication 17.1 (1979): 53-59.
Outlines a study undertaken to determine if any
significant differences existed between three methods
of teaching grammar and usage. Concludes that traditional classroom instruction seems to be the most effective method.

386. Thompson, Isabelle. "The Given/New Contract and Cohesion: Some Suggestions for Classroom Practice."
Journal of Technical Writing and Communication 15
(1985): 205-14.
Discusses the Given/New Contract and cohesion in
terms of how these concepts can be applied to the classroom. Describes problems that result from "violations"
of the Given/New contract and cohesion and suggests
strategies for correcting these problems.

3.7 Teaching Report Writing

387. Allen, J. W., Jr. "Documentation: A Functional Rationale." Journal of Technical Writing and Communication 8 (1978): 35-42.
Describes a method for documenting business and
technical reports.

388. Borden, Christopher. "Technical Writing and Post
Graduates at a Vocational Technical School." Technical
Writing Teacher 4 (1977): 98-101.
Describes seven steps of technical report writing
that are used in courses taught at the Southeast Regional
Vocational Technical School in South Easton Massachu-
setts.

389. Estrin, Herman A. "Six Innovative Methods of Teaching Technical Writing." Journal of Technical Writing
and Communication 9 (1979): 185-95.

Describes six teaching methods that helped students write better technical reports.

390. Fisher, Carol J. "Report Writing Skills." Instructor 89 (1979): 158-59.
Describes the skills necessary for good report writing.

391. Gisselman, Robert D. "A Nontraditional Report Writing Course: A Modest Proposal." Technical Writing Teacher 8 (1981): 62-65.
Discusses a conference-tutorial format used to teach an advanced report writing course. Includes descriptions of course content, evaluation techniques and conferencing.

392. Hall, John D. "A Continuing Case Approach: Report Writing Structure." ABCA Bulletin 41.3 (1978): 23-25.
Describes a structured approach to teaching report writing.

393. Jordan, Michael P. "New Directions in Teaching Technical Report Writing." Journal of Technical Writing and Communication 5 (1975): 199-205.
Describes the traditional method of teaching technical report writing and suggests ways to improve the traditional teaching method.

394. Koeppel, Mary Sue. "Making English Composition Skills Practical for the Students." Improving College and University Teaching 27.2 (1979): 72-74.
Describes a report writing program designed by a team consisting of an educator and a police training officer.

395. McGuire, Peter J. "Environmental Reports as Teaching Aids: Models, Audiences, Contexts." Technical Writing Teacher 8 (1981): 77-79.
Suggests using environmental reports to meet the varying needs found in a heterogeneous technical writing classroom. Explains how these reports serve as models, create a sense of audience, and provide contexts similar to those found in the workplace.

396. Minor, Dennis E. "An Integrated Technical Writing

Course." Technical Writing Teacher 3 (1975):   21-24.
States that students in technical writing courses
that include a major report project often do not understand the relevance of other writing assignments. Suggests that technical writing teachers can solve this
problem by "making as many of the assignments as possible contributory to the long report" (p. 21). Lists
various assignments and shows how these can be used
in conjunction with the formal report.

397. O'Keeffe, Katherine O'Brian, and Alan R. P. Journet.
"A Hypothetico-Deductive Model for Teaching the Research Paper." Journal of Technical Writing and Communication 15 (1985):   339-52.
Provides a model to be used to teach students how
to perform investigative research in order to produce
better research reports.

398. Perkins, Raymond. "A Dissepiment on Dissertating."
The Journal of Industrial Teacher Education 14.3 (1977):
36-37.
Examines five steps which promote efficiency in writing a research report.

399. Sherman, Dean. "The Binder Method: A Spatial,
Conceptual Approach to Teaching Business Report Writing." ABCA Bulletin 48.2 (1985):   26-28.
Describes a method used to teach students how to
produce business reports. Focuses on writing as a
"spatial" instead of a "chronological" process.

400. Warren, Thomas L. "Teaching Formal Technical Reports." Teaching English in the Two-Year College 8
(1982):   137-42.
Describes the merits of teaching formal technical
reports. Suggests using the technical report as an
assignment that incorporates many of the other assignments done by the student during the course. Provides suggested assignments and a timetable that teachers can use to teach the formal report in a sixteen-week
semester.

401. Washington, Gene. "Teaching the use of Declarative/
Procedural Information." Technical Communication 31
(1984):   39.

Describes a method for teaching students how to produce problem/solution reports.

402. Waterman, M. A., and J. F. Rissler. "Use of Scientific Research Reports to Develop Higher-Level Cognitive Skills." Journal of College Science Teaching 11 (1982): 336-40.
    Discusses how scientific reports can be used to improve writing skills of science majors.

403. Watt, James T., and Wade S. Hobbs. "Research and Report Writing: A Model of Their Interrelationships." ABCA Bulletin 42.3 (1979): 22-25.
    Describes a model that can be used by writers to define problems, research the problems, and write reports that present information concerning the problems.

3.8 Providing Strategies for Writing Various Documents

404. Ahern, Susan K. "Teaching the Technology--What Happens When the Writing Is Done." Technical Writing Teacher 8 (1981): 83-85.
    Describes the publishing process and the writer's role within this process to show technical writing teachers the importance of incorporating a unit on the publishing process into the technical writing classroom.

405. Bankston, Dorothy H. "Teaching the Detailed Description of a Mechanism." Technical Writing Teacher 2 (1975): 14-15.
    Describes how the author teaches the detailed description of a mechanism. States that this assignment 1) has "validity for a wide range of students," 2) is "interesting and challenging to write," and 3) is "practically impossible to plagiarize" (p. 14).

406. Beck, James P. "Toward a Model for Process--'Elucidation.'" Technical Writing Teacher 10 (1983): 108-13.
    Provides an eleven-step model for writing descriptions/instructions for "teaching-and-informing "all" about a complex sequence so that even a lay-novice reader can unmistakeably do it ... and/or also know it...." (p. 108).

407. Clark, Andrew K.   "The Abstract Way to Concrete
      Writing."   Journal of Technical Writing and Communica-
      tion 11 (1981):   131-38.
          Describes a unit on the teaching of how to write
      abstracts. Provides a sample syllabus and guidelines
      for writing abstracts.

408. Evans, Oliver H.   "Using Transmittal Correspondence
      as a Teaching Tool."   ABCA Bulletin 44.3 (1981):   18-
      19.
          Suggests that instructors should teach transmittal
      correspondence early in a technical writing course and
      then require each student to submit a transmittal memo
      with each assignment. Explains the importance of using
      transmittal memo assignments.

409. Haga, Enoch.   "1-2-3:   A New Way to Read and Write
      Letters."   The Journal of Business Education 52.5
      (1977):   229-30.
          Describes a method for teaching business letters.
      Gives sample assignments.

410. Halper, Cheryl A.   "Using Magazine Ads to Teach Sales
      Writing."   ABCA Bulletin 44.4 (1980):   22-23.
          Shows how magazine ads can be used to teach prin-
      ciples of writing effective sales letters. Provides guide-
      lines for producing sales letters.

411. Harder, Virgil E.   "Spelling: Can Teachers Train ALL
      Students to a 'Zero Defect' Level?"   ABCA Bulletin 46.1
      (1983):   19-21.
          Describes methods that the author tried to "train"
      his students to turn in application letters without mis-
      spelled words. Evaluates the methods and offers some
      suggestions for teachers.

412. Locker, Kitty O.   "Teaching Students to Write Ab-
      stracts."   Technical Writing Teacher 10 (1982):   17-20.
          Discusses a method for teaching abstracts. Examines
      various kinds of abstracts and presents guidelines for
      writing them. Includes sample assignments.

413. _____.   "Teaching Students to Write Abstracts."
      Workshop on the Teaching of Technical Writing Conven-
      tion. Carbondale, Illinois, 17-18 October 1980.

Describes a six-step strategy for teaching students how to write abstracts.

414. Masse, Roger E., and Patrick M. Kelly. "The Process of Teaching Description of a Process." Technical Writing Teacher 7 (1979): 15-18.
Describes how students at New Mexico State University are taught to write a description of a process by combining observation of the process actually being performed and writing assignments.

415. Mayer, Kenneth R. "Using a 'Lazlo Letter' in a Business Writing Class." ABCA Bulletin 45.2 (1982): 17-19.
Describes the use of sample business letters in the business writing classroom.

416. Nash, Diane. "From Writing for Grades to Writing for Pay." Media and Methods 15.3 (1978): 43-44.
Examines the advantages of having students write for publication in magazines.

417. Plung, Daniel L. "Writing the Persuasive Business Letter." The Journal of Business Communication 17.3 (1980): 45-49.
Draws from John Dewey's concept of the "motivated sequence" to design a strategy for producing persuasive business letters.

418. Poe, Roy W. "Teaching Communication and Management Skills with Business Letter and Report Writing." Business Education World 58.3 (1978): 6-7, 25.
Examines the use of situational approach to teach business letter and report writing. Stresses the importance of developing good communication skills.

419. Roberts, David. "Teaching Abstracts in Technical Writing: Early and Often." Technical Writing Teacher 10 (1982): 12-16.
Stresses the importance of writing abstracts. Presents a method for teaching abstract writing.

420. Sachs, Harley L. "Technical Writing Students as Audiences." Technical Writing Teacher 7 (1980): 122-24.

Describes an exercise used to teach students about
audience awareness.

421. Sherrard, Carol A.   "The Psychology of Summary-
      Writing." Journal of Technical Writing and Communica-
      tion 15 (1985):   247-58.
         Explains how findings from research conducted in
      the area of cognitive psychology can be used in the
      classroom to teach students how to write effective sum-
      maries.

422. Walsh, E. Michael.   "Teaching the Letter of Applica-
      tion." College Composition and Communication 28 (1977):
      374-76.
         Suggests that the letter of application should be
      taught by a method that emphasizes "self-analysis,"
      "job analysis," and "self-marketing." Provides a class-
      by-class discussion of how to teach the letter of ap-
      plication.

423. Waltman, John L., and Steven P. Golen.   "Resolving
      Problems of Invention in Sales Letters." ABCA Bulletin
      47.2 (1984):   19-20.
         Discusses the use of application letters and sales
      letters to help students learn how to tailor information
      to satisfy audiences and purposes.

424. Weiss, Timothy.   "Sums Are Not Set on Erasmus (or,
      A Three-Step Method for Teaching Technical Descrip-
      tions)." The Journal of Business Communication 22.4
      (1985):   51-57.
         Offers a new approach to teaching technical descrip-
      tions to make technical description assignments inter-
      esting and purposeful for students.

425. White, Leigh Cree.   "A Pattern for Publication." Jour-
      nal of Technical Writing and Communication 11 (1981):
      45-56.
         Describes a "publication model" used to help students
      realize that writing is "just a part of the freelance
      process" (p. 45).

426. Wunsch, Alan P.   "Letter Writing Seminar for Word
      Processing Principals." The Journal of Business Edu-
      cation 53 (1978):   263-64.

Describes a letter writing seminar taught at the American Express Company in Phoenix Arizona.

3.9 Applying Various Approaches

427.   Baker, William H., and Nadine T. Ashby. "Teaching Business Writing by the Spiral Method." The Journal of Business Communication 14.3 (1977): 13-22.
Discusses several factors that affect learning. Uses these factors to develop a "spiral method" to teaching. Presents a step-by-step explanation of how to use the spiral method. Concludes that the spiral method is flexible, so it meets the needs of many students.

428.   Benton, Sharon K. "Techniques for the Technical Writer." Performance and Instruction 24.5 (1985): 4-5.
Discusses strategies for writing effective materials to be used in training.

429.   Brostoff, Anita. "Applications of the Functional Writing Model in Technical and Professional Writing." Conference on English Education. Minneapolis, Minnesota. 16-18 March 1978.
Explains how a "functional writing model" can be used in the teaching of technical writing.

430.   Bruckmann, C. G. "A Systems Model of the Communication Process." Journal of Technical Writing and Communication 8 (1978): 321-42.
Describes a model that "consists essentially of a set of systems and subsystems which can be combined to represent any human communication situation" (p. 322). Explains how the model works and gives examples to show how the model applies to various communication situations.

431.   Casstevens, E. Reber. "An Approach to Communication Model Building." The Journal of Business Communication 16.3 (1979): 31-40.
Argues that instead of trying to develop new communication models, we can change the basic communication model, consisting of sender, channel and receiver, to fit any need we have as users of the model. Gives examples to show how the basic model can be refined to meet different needs.

432. Debs, Mary Beth, and Lia Brillhart. "Technical Lis-
      tening/Technical Writing." Technical Writing Teacher
      8 (1981): 83-84.
          Explains that "listening" is a skill that professionals
      need. Describes how at Triton, an Illinois community
      college, listening and writing skills "are taught in con-
      junction" through the use of guest speakers and writing
      assignments (p. 83).

433. Drew, Sara. "Tailored Technical Writing." Teaching
      English in the Two-Year College 7 (1980):   55-56.
          Describes a method for effectively teaching technical
      writing in a classroom that consists of students from a
      variety of majors. Explains how to teach the students
      by using individualized modules designed for the stu-
      dents' majors.

434. Falcione, Raymond L. "Some Instructional Strategies
      in the Teaching of Organizational Communication." The
      Journal of Business Communication 14.2 (1977):   21-34.
          Provides an instructional paradigm that can be used
      in "any learning environment." Discusses four strate-
      gies that can be used to teach organizational communi-
      cation.

435. Flatley, Marie E., and Maurice L. Crawford. "Managing
      the Large Business Communication Class." ABCA Bul-
      letin 46.1 (1983):   7-9.
          Discusses the "nature" of the large business class.
      Offers suggestions for how to plan, teach, and manage
      the large class.

436. Fraley, Lawrence E., and Ernest A. Vargas. "Modular
      Instruction: Its Structure, Operation, and Implications."
      Journal of Technical Writing and Communication 6 (1976):
      57-70.
          Describes how modules organize concept into units
      that individuals can work on alone in order to achieve
      specified instructional activities. Explains how to set
      up modules and how to evaluate the results of modular
      instruction. Outlines the advantages of a modular cur-
      riculum.

437. Freese, Robert L., and others. "'You Want Me to Do
      What?'" Technical Writing Teacher 6 (1979): 108.

Describes a method for teaching spelling that is used by Technical Communications instructors at the South Dakota School of Mines and Technology.

438. Giovannini, Mary, and F. Milton Miller. "Teaching Business Communications: A Comparison of Teaching by the Traditional Writing and the Word Processing Methods." Journal in the Studies of Technical Careers 6.1 (1984): 21-28.
Reports the findings of a study conducted to compare the effectiveness of traditional and word processing methods for teaching business communication.

439. Golen, Steven, and others. "How to Teach Students to Improve Their Creativity in a Basic Business Communication Class." The Journal of Business Communication 20.3 (1983): 47-57.
Presents a unit that teachers can use to promote creativity in their business communication classrooms. Includes an outline for the unit, sample assignments, and a list of resource materials.

440. Golen, Steven; Alan Burns; and James W. Gentry. "An Analysis of Communication Barriers in Five Methods of Teaching Business Subjects." The Journal of Business Communication 21.3 (1984): 45-52.
Describes a study to determine communication barriers that hinder the effectiveness of five teaching methods frequently used in business and technical writing courses.

441. Halpern, Jeanne W. "How to Start a Business Writing Lab." ABCA Bulletin 44.4 (1981): 9-14.
Explains the merits of using a business writing lab as a "supplement to classroom instruction" (p. 9). Provides guidelines for setting up such a lab, basing suggested guidelines on the experience of setting up such a lab at Purdue University.

442. Hamilton, David. "Writing Science." College English 40 (1978): 32-40.
Explains how to teach writing as "an integral act of science" in order to show the role of creativity in science.

443. Hart, Maxine Barton. "An Experimental Study in Teach-
      ing Business Communications Using Two Different Ap-
      proaches: Theory and Application Approach vs. Writing
      Approach." The Journal of Business Communication
      17-1 (1980): 13-25.
          Outlines a study undertaken to determine whether
      the traditional writing approach or the theory/applica-
      tion approach is more effective in teaching business
      communication. Lists several ways that the addition
      of theory into the classroom was advantageous.

444. Held, Julie Stusrud. "Teaching Writers How to Write:
      What Works?" Technical Communication 30.2 (1983):
      17-19.
          Reviews literature on the teaching of writing and
      discusses what this literature has to offer teachers of
      technical writing.

445. Johnson, Betty S., and Jeannette W. Vaughn. "Time
      Management and Communication: Integrating Skills for
      Higher Productivity." Journal of Technical Writing and
      Communication 15 (1985): 267-77.
          Stresses the importance of using time efficiently.
      Suggests ways to help students learn to better manage
      time to achieve greater productivity.

446. Kaltenbach, Joan C. "The Effects of Two Teaching
      Methodologies on the Performance and Attitudes of
      Students in a Technical Report-Writing Course." Jour-
      nal of Technical Writing and Communication 10 (1980):
      323-27.
          Discusses the results of an experiment conducted to
      contrast two teaching methodologies: the traditional
      lecture approach and the individual conference approach.
      Describes how the experiment was conducted and con-
      cludes that the individual conference approach produced
      better student performance.

447. Locker, Kitty. "Teaching with Transparencies." ABCA
      Bulletin 40.3 (1977): 21-27.
          Explains how transparencies can be used to teach
      effective use of content and format in business docu-
      ments.

448. Lynch, David. "Communication: Help Your Students

to Express Ideas More Concretely." Business Education
Forum 34.7 (1980): 17-19.
    Discusses the importance of being specific when ex-
pressing ideas in business writing. Suggests that
teachers can encourage students to be more observant
of detail by having students take notes on events that
they observe and asking students to incorporate their
notes into their writings.

449. Madigan, Chris. "Improving Writing Assignments with
Communication Theory." College Composition and Communication 36 (1985): 183-90.
    Explains how to use Jakobson's communication model
to design better writing assignments.

450. Mancuso, Joseph C. "The Importance of Individual
Student Conferences in Technical Writing Courses."
Journal of Technical Writing and Communication 14
(1984): 203-06.
    Presents the advantages derived by both students
and instructors from the use of individual student conferences.

451. Marcus, Kathleen. "Résumé Writing: A New Approach."
Journal of College Placement 40 (1979): 61-65.
    Explains how a slide/tape presentation and a work-
book are used to teach students about the job search
and resume writing.

452. Matulich, Loretta. "Contract Learning in the Traditional
Technical Writing Class." Technical Writing Teacher
11 (1984): 110-15.
    Describes a "contract learning system" for a techni-
cal writing class. Provides models for designing contracts, writing evaluations of contracts, and writing
self-evaluation guidelines. Finds the contract learning
system to be an effective teaching method.

453. Miller, Carolyn R. "Technical Writing Textbooks: Current Alternatives in Teaching." Technical Communication 31.4 (1984): 35-38.
    Examines various approaches to teaching technical
writing as reflected in textbooks.

454. Moldstad, John A. "Selective Review of Research Studies

Showing Media Effectiveness: A Primer for Media Di-
rectors." Journal of Technical Writing and Communication 5 (1975): 323-41.
  Evaluates the effectiveness of using various media
for instructional purposes.

455. Morton, Gerald W. "Bridging the Gap Between the
      Freshman Essay and the Technical Report." Technical
      Writing Teacher 10 (1983): 134-39.
        Describes how the technical writing teacher can inte-
      grate concepts "taught in freshman composition--thesis
      writing, using basic expository patterns, organizing an
      essay, using transitions--" (p. 134) into the technical
      writing classroom by pointing out the similarities be-
      tween the freshman essay and the technical report.

456. Pace, R. Wayne. "An Experiential Approach to Teach-
      ing Organizational Communication." The Journal of
      Business Communication 14.4 (1977): 37-47.
        Discusses the merits of experiential learning and
      describes concepts that are central to organizational
      communication. Provides a list of exercises that teach-
      ers can use to emphasize experiential learning while
      instructing students in the concepts of organizational
      communication.

457. Pieper, Gail W. "Scientific Documentation." Technical
      Writing Teacher 7 (1980): 71-73.
        Describes three fifty-minute sessions that focus on
      three types of scientific documentation and the formats
      used for each.

458. Ponthieu, J. F., and James T. Watt. "Broadening the
      Base: A New Approach to Teaching Business Writing."
      ABCA Bulletin 39.3 (1976): 9-12.
        Examines the use of team teaching in the business
      writing classroom.

459. Pundsack, Victoria. "Technical Writing and Native
      Good Judgment." Technical Writing Teacher 5 (1978):
      85-87.
        Discusses methods for teaching students audience
      awareness.

460. Ramsden, Patricia A. "Workshops in Technical Com-

position." Technical Writing Teacher 7 (1980):   69-70.
Finds that the traditional lecture/discussion method
of teaching technical writing is not effective. Suggests
a workshop teaching method which combines lectures
and workshops to provide students with more individ-
ualized instruction.

461. Rosner, Mary, and Terri Paul. "Using Sentence Combining in Technical Writing Classes." Conference on
CCC. Dallas, Texas, 26-28 March 1981.
Describes various sentence combining exercises that
can be used in the technical writing class.

462. Rothwell, William J. "A Career Planning Questionnaire."
ABCA Bulletin 47.2 (1984):   15.
Presents a questionnaire which allows students to
assess their career goals so that teachers can better
design units on the job search to meet student's needs.

463. _____. "The Business Writing Instructor as Coun-
selor." ABCA Bulletin 47.3 (1984):   49-52.
Examines three counseling theories and explains how
the theories are applicable to the business writing classroom.

464. Roundy, Nancy. "Revision Pedagogy in Technical Writing." Conference on CCC. New York, New York, 29-
31 March 1984.
Discusses what needs to be done to teach revision
in the technical writing classroom.

465. _____. "Team-Teaching Technical Writing: A Practical Approach to a Practical Discipline." Technical
Writing Teacher 9 (1981):   23-24.
Describes team-teaching approaches used at the University of Oklahoma to tie instruction given in the
technical writing classroom to the students' major fields
of study. Finds that team-teaching motivates students
by providing them with writing assignments similar to
those they will later do on the job.

466. _____. "The Heuristics of Pedagogy: Approaches
to Teaching Technical Writing." Conference on CCC.
Minneapolis, Minnesota, 21-23 March 1985.
Describes a framework for evaluating pedagogical
materials.

467. Samuels, Marilyn Schauer. "Scientific Logic: A
      Reader-Oriented Approach to Technical Writing." Jour-
      nal of Technical Writing and Communication 12 (1982):
      307-28.
          Describes three different analysis of the "thinking
      and writing process": 1) Martin S. Peterson's Mountain/
      Pyramid, 2) Thomas P. Johnson's Inverted-Pyramid and
      3) Barbara Minto's Pyramid Principle. Shows how these
      approaches can be combined and used by the technical
      writing teacher to teach audience analysis.

468. Santelmann, Patricia Kelly. "Teaching Technical Writ-
      ing: Focusing on Process." Conference on CCC. Min-
      neapolis, Minnesota, 21-23 March 1985.
          Describes how to teach technical writing students all
      areas they must consider when writing technical docu-
      ments.

469. Schenck, Eleanor M. "Technical Writers, Readers, and
      Context Clues." Journal of Technical Writing and Com-
      munication 10 (1980): 189-94.
          Examines the importance that vocabulary study has
      for technical writers. Describes a study undertaken
      to determine how context clues are used with regard
      to the intended audiences. Explains how a system used
      to teach vocabulary to readers can be used to teach
      technical writers.

470. Selzer, Jack. "Resources for Teachers of Technical
      Writing." Technical Communication 31.4 (1984): 39-44.
          Presents various resources that are available to
      help better prepare teachers of technical writing.

471. Shelby, Annette. "A Teaching Module: Corporate Ad-
      vocacy." ABCA Bulletin 47.2 (1984): 16-19.
          Explains the importance of teaching students how to
      handle media attacks of businesses. Outlines a teach-
      ing unit on corporate advocacy.

472. Sides, Charles H. "Heuristics or Prescription: Synthe-
      sis Rather than Choice." Journal of Technical Writing
      and Communication 11 (1981): 115-20.
          Offers a "middle-ground" approach to solving the
      issue of whether prescription or heuristics should be
      used in the teaching of technical writing. Suggests

using heuristics "as prewriting, discovery, and invention procedures, as provisions for solving communication problems" and suggests that prescription must be used to provide students with guidelines for producing technical documents. Provides a brief annotated bibliography on heuristics and prescription.

473. Sills, Caryl Klein. "Designing Forms for the Electronic Age." Exercise Exchange 29.2 (1984): 36-40.
Discusses an assignment used in business and technical writing classes to teach ways to design forms.

474. Smith, Barbara. "Experiential Learning in Technical Writing Courses." Technical Writing Teacher 7 (1980): 102-07.
Discusses the importance of using experiential learning in the technical writing classroom. Provides examples of students' learning from experience.

475. "Some Suggestions for Having Guest Speakers in a Technical Writing Course." Technical Writing Teacher 7 (1980): 64-65.
Provides suggestions for effectively selecting and incorporating guest speakers in the technical writing classroom.

476. Stevenson, Dwight W., and Peter R. Klaver. "Problem Definition for Problem Solvers: Applying Rhetorical Theory in Teaching Technical Writing." Conference on CCC. Denver, Colorado, 30 March-1 April 1978.
Examines the use of "tagmemic rhetoric" to improve technical writing students' problem-solving skills.

477. Strauss, Andre. "Does Teaching Involve Informing?" Journal of Technical Writing and Communication 11 (1981): 1-8.
Discusses the nature of "teaching." Finds that teachers should help students learn to analyze information.

478. Switzler, Al. "Reality Therapy and the Technical Writing Classroom." Technical Writing Teacher 4 (1977): 89-93.
Explains what reality therapy has to offer the teacher of technical writing.

479.   Waltman, John L.   "Entropy and Business Communica-
       tion."   The Journal of Business Communication 21.1
       (1984):   63-80.
           Describes the development of the concept of entropy.
       Examines the application of entropy to problems in busi-
       ness communication.

480.   Wolf, Morris Philip.   "S. I. Hayakawa's Scholarship
       and the Teaching of Business Communication."   ABCA
       Bulletin 48.1 (1985):   5-7.
           Explains how Hayakawa's interpretation of Korzybski's
       General Semantics Theory can be adapted for practical
       use in teaching and researching business communica-
       tion.

481.   Wright, Patricia.   "Behavioral Research and the Tech-
       nical Communicator."   Technical Communication 25.2
       (1978):   6-12.
           Describes three human behavioral experiments and
       explains how findings from such experiments are rele-
       vant for technical writers.

482.   Wyld, Lionel D.   "Beyond Written Communication: A/V
       in the Classroom.   Journal of Technical Writing and
       Communication 13 (1983):   229-33.
           Examines the instructional uses of audiovisual ma-
       terials.

483.   Yoder, Albert C.   "Technical English and the Future."
       Journal of Technical Writing and Communication 8 (1978):
       359-62.
           Discusses two methods for teaching technical English.
       Suggests that combining these two methods will improve
       instruction.

• 4. GRADING TECHNIQUES •

484. Brogan, Katherine M., and James D. Brogan. "Yet Another Ethical Problem in Technical Writing." Annual Meeting of College English Association. (13th, Houston, TX, April 15-17, 1982).
Discusses the problems of plagiarism in technical reports.

485. Ceccio, Joseph F. "Checkmark Grading and the Quarter System." ABCA Bulletin 39.3 (1976): 7-9.
Explains the effectiveness of using behaviorial grading in the business writing classroom.

486. Connor, Jennifer J. "Submissions in the Technical Writing Course: Toward Ensuring Their Originality." English Quarterly 18.2 (1985): 53-56.
Suggests ways for decreasing the instances of plagiarism in the technical writing classroom.

487. Dudenhefer, John Paul. "An Experiment in Grading the Writing of Technical Students in Developmental English." Journal of Technical Writing and Communication 7 (1977): 121-128.
Describes a study to determine if a student's revising of a writing sample before the teacher's grading of that sample improved the student's writing.

488. Dulek, Ron, and Annette Shelby. "Varying Evaluative Criteria: A Factor in Differential Grading: The Journal of Business Communication 18.2 (1981): 41-50.
Describes a study conducted to determine what causes "significant disparities in evaluation" (p. 42).

489. Feinberg, Susan. "Evaluation of Business Communication Techniques." The Journal of Business Communication, 17.4 (1980): 43-50.
Describes an experiment conducted to determine the effectiveness of holistic evaluation.

490.   Hollis, L. Y.   "Translating Competencies into Perfor-
       mance Measures for the Evaluation of Teaching."   Amer-
       ican Association of Colleges for Teacher Education Con-
       vention, Chicago, 21-24 February 1978.
            Explains how to devise instruments that teachers
       can use to evaluate student performance.

491.   Houston, Linda.   "Grading--The Student's Option."
       The Technical Writing Teacher 11.1 Fall 1983, 21-22.
            Explains how optional writing assignments can be
       used to determine if technical writing students can
       "transfer their skills" when working independently of
       the teacher.

492.   Lally, Joan M.   "Are Objective Pretests Useful?"   ABCA
       Bulletin 44.4 (1980):   26-27.
            Finds that objective pretests are useful in determin-
       ing how students will perform in the business communi-
       cation classroom.

493.   Lupul, Ruth.   "In Praise of the Point System."   ABCA
       Bulletin 46.1, March 1983, 12-13.
            Discusses the advantages to using the point grading
       system instead of letter grading.   Finds that business
       students respond well to the point grading system.

494.   Miller, Carol.   "Commentary That Works for the Learn-
       ing Business Writer."   The Journal of Business Com-
       munication 22.4 (1985):   43-50.
            Explains what constitutes productive and nonproduc-
       tive feedback from instructors.   Shows how the use of
       productive feedback can be used to help improve stu-
       dents writing in business and technical communication
       classroom.

495.   Murphree, Carolyn T.   "A Consideration of Competency
       Testing."   ABCA Bulletin 48.2 (1985):   24-25.
            Presents the steps involved in choosing and adminis-
       tering a competency test.

496.   Newsom, Doug, and Tom Siegried.   "Public Relations
       Writing: Setting Goals and Objectives for Students and
       Evaluating Their Work."   Association for Education in
       Journalism Conference, East Lansing, Michigan, 8-11
       August 1981.

Discusses a systematic method for evaluating students written work.

497. Ober, Scot. "The Influence of Selected Variables on the Grading of Student-Written Letters." ABCA Bulletin 47.1, 7-11.
Describes a study to determine if the grades teachers assign are influenced by "format."

498. Potvin, Janet H. "A Systems Approach for Teaching and Evaluating Technical Writing." Technical Writing Teacher 10.3, Spring 1983, 210-17.
"Describes a series of checklists" that help technical writing students as they plan and write papers and serve as evaluative tools for technical writing teachers. Provides examples of such checklists. Also evaluates the effectiveness of using checklists "to improve student writing" (p. 210).

499. Rentry, Kathryn C. "Some Discouraging Words About Checkmark Grading." ABCA Bulletin 48.2, 20-23.
Examines some of the ways that checkmark grading is ineffective. Describes the author's experiences with using checkmark grading.

500. Ross, Donald, Jr. "A Brief Note on How Writing Errors Are Judged." Journal of Technical Writing and Communication 11 (1981): 163-73.
Describes a study to determine how students respond to errors in written documents.

501. Rossi, Michael J. "A Grading System That Credits Students for What They Do Well." ABCA Bulletin 43.3 (1980): 12-15.
Outlines a grading system that is both objective and flexible.

502. Rutter, Russell. "Starting to Write by Rewriting: A Unit on Technical Editing and Revision." Technical Writing Teacher 8 (1980): 22-26.
Describes a diagnostic unit that teachers can use to determine the "needs" of each particular group of students as well as tell "the students what technical writing is and what resources and common concerns they bring to the study and practice of it" (p. 22).

503.   Stratton, Charles R.   "The Electric Report Card: A
       Followup on Cassett Grading." Journal of Technical
       Writing and Communication 5 (1975):   17-22.
           Describes how cassette grading is used to grade
       students papers in Technical and Engineering Report
       Writing classes at the University of Idaho.

504.   Sullivan, Patricia A., and David D. Roberts.   "A Case
       for Diagnosis in Technical Writing." Journal of Tech-
       nical Writing and Communication 11 (1981):   35-43.
           Describes a diagnostic tool that can be used to
       evaluate technical writing classes.

505.   Swenson, Dan H., and others.   "Reducing the Number
       of Teacher-Graded Papers in the Teaching of Informa-
       tional Business Writing." American Business Communi-
       cation Association Conference, Phoenix, 14-17 October
       1981.
           Describes a study to determine the effect of reduc-
       ing the number of teacher-graded assignments.

506.   Tebeaux, Elizabeth.   "Grade Report Writing with a
       Check Sheet." Technical Writing Teacher 7 (1980):
       66-68.
           Describes a check sheet devised to improve methods
       for grading technical reports written by students in
       technical writing classes. Discusses the advantages of
       using the check sheet. Includes a sample check sheet.

507.   Throop, David P.; and Daphne A. Jameson.   "Behav-
       ioral Grading: An Approach Worth Trying." ABCA
       Bulletin 39.3 (1976):   3-5.
           Explains how behaviorial grading can be used ef-
       fectively in the technical writing classroom.

508.   Warren, Thomas L.   "Objective grading standards in
       Technical Writing." Technical Writing Teacher 4 (1976):
       29-31.
           Provides a sample grading form used to evaluate
       student reports.

509.   White, Kathryn F., and Betty S. Johnson.   "Evaluation
       Criteria Used in Business Communication Courses Taught
       in College of Business Affiliated with AACSB." ABCA
       Bulletin 47.3, 39-42.

Describes a study to determine what business communication instructors used for evaluating writing assignments and determining course grades.

510.   Wilkinson, Dorothy C. "Evidence That Others Do Not Agree with Your Grading of Letters." ABCA Bulletin 42.3 (1979): 29-30.
Describes a study that showed how teachers varied in their grading of business letters.

PART II:

ASPECTS OF BUSINESS AND TECHNICAL
COMMUNICATION

5. THE NATURE OF TECHNICAL WRITERS
AND THE WRITING PROCESS

511. Booher, E. Kathleen, and Wilam Davidson. "Eleven Myths About Writing--And How Trainers Can Debunk Them." IEEE--Transactions on Professional Communications PC-25 (1982): 133-35.
Describes eleven myths about writing and explains why these myths should not be accepted as truths. Suggests that "a writing trainer or consultant should be a writer and a trainer" (p. 135).

512. Brucher, Richard T. "Definition: The First Step in the Thinking/Writing Process." Journal of Technical Writing and Communication 10 (1980): 249-55.
Explains how to use definition to help students better understand the writing process.

513. Cederborg, Gibson A. "Tinker, Tailor, Technical Communicator." Journal of Technical Writing and Communication 14 (1984): 217-22.
Uses the "Job and Talent Matrix" to determine characteristics of technical writers and editors.

514. Cory, W. A., Jr. "You Can't Write!" The Science Teacher 49.6 (1982): 51-53.
Examines the importance of good writing skills to scientists. Suggests ways to improve writing skills.

515. De Gise, Robert F. "A Systems Approach to Business Writing." IEEE--Transaction on Professional Communications PC-23 (1980): 77-78.
Describes a four-step approach to make the writing process less difficult.

516. Douglas, George H. "The Informational Sketch--A Major Vice of the Technical Writer." Journal of Technical Writing and Communication 12 (1982): 7-13.

Examines causes of writing that contains no substance. Suggests ways to produce writing with substance.

517. Fagan, Bon. "Executive Writing and Internal Revision."
      English Quarterly 18 (1985): 69-74.
      Discusses similarities between an executive's writing
      processes and those of students.

518. Garver, Eugene. "Demystifying Classical Rhetoric."
      ABCA Bulletin 47.3 (1984): 24-28.
      Explains how classical invention techniques are useful for student writers of today.

519. Gilsdorf, Jeanette W.   "The Modified Sentence Outlines:
      A Timesaving Report Format." ABCA Bulletin 45.1
      (1982): 15-18.
      Explains how a modified sentence outline can be used
      in report writing.

520. Gould, Jay R. "Support Material Adds to the Bare
      Bones." Journal of Technical Writing and Communication 6 (1976): 315-25.
      Describes various techniques that can be used to
      "support" more general statements presented in documents.

521. Gross, Alan G. "The Form of the Experimental Paper:
      A Realization of the Myth of Induction." Journal of
      Technical Writing and Communication 15 (1985): 15-26.
      Explains how the organization of experimental papers
      reflect the effort to cling to scientific inductive reasoning.

522. Guthrie, John T. "Research: Rhetorical Awareness."
      Journal of Reading 28.1 (1984): 92-93.
      Discusses research conducted to determine nonacademic writers' cognitive processes as they compose.

523. Halpern, Jeanne W. "Effects of Dictation/Word Processing System on Teaching Writing." Conference on CCC,
      San Francisco, California, 18-20 March 1982.
      Describes a study to determine how new technological developments affect writer's composing processes.

524. Hays, Robert. "Model Outlines Can Make Routine Writing Easier." Technical Communication 29 (1982): 4-8.
Discusses eleven models that writers can use to make the writing task easier.

525. Huber, Carol E. "The Logical Art of Writing Useful Comparisons." IEEE--Transactions on Professional Communications PC-28 (1985): 27-28.
Explains how to use logical reasoning to generate effective comparisons that can be used when explaining concepts to audiences.

526. Hulbert, Jack E. "Effective Business Writing." IEEE--Transactions on Professional Communications PC-23 (1980): 128-29.
Discusses the components of an effective writer.

527. Kallendorf, Craig, and Carol Kallendorf. "A New Topical System for Corporate Speechwriting." The Journal of Business Communication 21.2 (1984): 3-14.
Draws from classical rhetoric to devise a list of "topics" that business communicators can use to generate arguments.

528. Kinyon, Jeannette E. "I Am No Wolf, and You Are Not Rabbits." Journal of Technical Writing and Communication 6 (1976): 195-201.
Compares the writing process to juggling. States that learning to write well involves thinking and practicing. Explains what the relationship between teachers and students should be.

529. Knoblauch, C. H. "Intentionality in the Writing Process: A Case Study." College Composition and Communication 31 (1980): 153-59.
Distinguishes between "generic" purposes in writing, which are "ideal" and "simple," and "operational" purposes in writing, which "are specific to real situations and often quite complex" (p. 154). Discusses information obtained by interviewing 250 executives of a large consulting firm and by evaluating numerous proposals written by executives in the firm. Uses this information to support the argument that operational purposes must be accounted for in any comprehensive theory of discourse.

530. Lay, Mary M. "Teaching Narrative in Technical Writing
      Classes." Technical Writing Teacher 9 (1982): 156-58.
          Shows how the use of narrative techniques can im-
      prove students analyses of writing situations.

531. _____. "The Enthymeme: A Persuasive Strategy in
      Technical Writing." Technical Communication 31.2
      (1984): 12-15.
          Discusses how the enthymeme can be used to produce
      effective communication in business and industry.

532. Locker, Kitty. "Patterns of Organization for Business
      Writing." Journal of Business Communication 14.3
      (1977): 35-46.
          Presents various patterns that can be used to or-
      ganize information in a document.

533. Marcus, Stephen, and Sheridan Blau. "Not Seeing Is
      Relieving: Invisible Writing with Computers." IEEE--
      Transactions on Professional Communications PC-27
      (1984): 69-72.
          Describes an experiment to determine the importance
      of being able to view the text as it is composed.

534. Masse, Roger E. "Theory and Practice of Writing Proc-
      esses for Technical Writers." IEEE--Transactions on
      Professional Communications PC-27 (1984): 185-92.
          Discusses an approach for better understanding the
      writing process.

535. McKee, Blain K. "Types of Outlines Used by Techni-
      cal Writers." Journal of English Teaching Techniques
      7.4 (1975): 30-36.
          Describes a survey conducted to determine if and
      how writers use outlines when writing.

536. McLeod, Susan H. "Use Piaget's Theories to Teach
      Technical Description." Technical Communication 32.2
      (1985): 47.
          Describes two assignments that requires students to
      use principles from Paiget's "theory of intellectual de-
      velopment" to help them think through and organize
      writing tasks.

537. Miller, Ryle L., Jr. "Translate Your Thoughts into

Good Technical Writing." IEEE--Transactions on Professional Communications PC-24 (1981): 73-74.

States that thoughts should be ordered as follows: 1) situation, 2) problem, 3) resolution, and 4) information.

538. Odell, Lee, and Dixie Goswami. "Writing in a Non-Academic Setting." Research in the Teaching of English 16 (1982): 201-23.

Examines writers in the business world to determine how important they consider rhetorical principles to be to the writing process.

539. Peterson, Robert J. "Brain-Flow Writing." Presented at the American Business Communication Association Southeast Convention. Hammond, Louisiana, 5-7 April 1984.

Suggests a method for helping writers overcome writer's block.

540. Plung, Daniel L. "The Advantages of Sentence Outlining." Technical Communication 29 (1982): 8-11.

Shows how using sentence outlines can help technical writers.

541. Pomerenke, Paula J. "Rewriting and Peer Evaluation: A Positive Approach to Business Writing Classes." ABCA Bulletin 47.3 (1984): 33-36.

Explains why rewriting is important in improving students' writing skills. Describes an experiment the author conducted to examine students' rewriting processes.

542. Radice, F. W. "Using the Cloze Procedure as a Teaching Technique." English Language Teaching Journal 32 (1978): 201-04.

Explains how to use the cloze procedure to teach the writing process.

543. Reed, W. Michael, and others. "The Effects of Writing Ability and Mode of Discourse on Cognitive Capacity Engagement." Research in the Teaching of English 19 (1985): 283-97.

Describes a study to determine the effects that level of writing ability and discourse mode has on the degree of a writer's involvement in composing processes.

544. Root, Robert L. "Style and Self: The Emergence of
      Voice." Journal of Teaching Writing 4.1 (1985): 77-85.
          Discusses how and when writers develop a "voice"
      in their writing.

545. Roundy, Nancy. "The Composing Process and Pedagogy;
      or A 'How-To' on Describing an Item." Technical Writ-
      ing Teacher 9 (1982): 159-62.
          Discusses an exercise that can be used to teach stu-
      dents about the composing process.

546. Schumacher, Gary M., and others. "Cognitive Activi-
      ties of Beginning and Advanced College Writers: A
      Pausal Analysis." Research in the Teaching of English
      18 (1984): 169-87.
          Discusses a study to determine what mental proc-
      esses occur when students pause as they are writing.

547. Selzer, Jack. "The Composing Process of an Engineer."
      College Composition and Communication 34 (1983): 178-
      87.
          Examines the composing process of an engineer to
      determine if the engineer's actual composing process
      matches general beliefs concerning the composing proc-
      ess.

548. Sharbrough, William C. "Increasing Student Involve-
      ment: Using Nominal Grouping to Generate Report
      Topics." ABCA Bulletin 45.2 (1982): 44-45.
          Describes a three-step process that will help stu-
      dents develop formal report topics.

549. Shear, Marie. "Fixing Rotten Writing: A Cameo Case
      History." The Journal of Business Communication
      18.2 (1981): 5-14.
          Describes the author's findings from an evaluation
      made on her own revision processes.

550. Smith, Frank. "Myth of Writing." IEEE--Transactions
      Professional Communications PC-27 (1984): 101-04.
          Discusses twenty-two myths concerning writing.

551. "The Practical Writer." IEEE--Transactions on Profes-
      sional Communications PC-24 (1981): 63-65.
          Discusses the qualities of a good writer.

552. Tripp, Janice A. "Invention in Technical Writing."
Conference on CCC. Denver, Colorado, 30 March-1
April 1978.
  Describes the importance of using invention activi-
ties in technical writing courses.

553. Washington, Gene. "Information as Visual Properties:
A Heuristic for Technical Writers." Technical Writing
Teacher 7 (1980): 94-101.
  Describes a heuristic to help students generate in-
formation concerning their subjects of writing. Ex-
plains how to teach this heuristic to students.

554. Wason, Peter C. "Conformity and Commitment in Writing." Visible Language 14 (1980): 351-63.
  Discusses the importance of writing and revising to
produce an effective document which says exactly what
the writer intended for it to say.

555. Winterowd, W. Ross. "Writing About Writing." Exercise Exchange 30.2 (1985): 26-27.
  Describes an assignment that asks students to write
a paper on their composing processes.

556. Wright, Patricia. "Five Skills Technical Writers Need."
IEEE--Transactions on Professional Communications
PC-24 (1981): 10-16.
  Discusses five skills that writers should possess.

557. Zeidner, Martin A. "Creative Technical Writing, the
Key to Successful Grantspersonship." Journal of Technical Writing and Communication 15 (1985): 69-73.
  Discusses the roles that creative thinking and news
services have in the process of obtaining research
grants.

558. Alderson, John W., and Robert D. Hay. "The Effect
      of Order of Information in a Questionnaire Cover Letter
      Situation." The Journal of Business Communication
      13.2 (1976): 11-15.
         Describes an experiment conducted to determine the
      effect that organization of information in a document
      has on the effectiveness of the document's message.

559. Benson, Philippa J. "Writing Visually: Design Con-
      siderations in Technical Publications." Techncial Com-
      munication 32.4 (1985): 35-39.
         Offers suggestions for effectively designing docu-
      ments to increase ease of readability and use.

560. Benton, Stephen L., and others. "Cognitive Capacity
      Differences Among Writers." Journal of Educational
      Psychology 76 (1984): 820-34.
         Describes studies conducted to compare the cognitive
      processes of good and poor writers.

561. Bogert, Judith. "In Defense of the Fog Index." ABCA
      Bulletin 48.2 (1985): 9-12.
         Defends readability indexes by arguing that the
      problems we find with readability formulas stem from
      our use of them for purposes other than those the in-
      dexes were designed to serve.

562. Bond, Sandra J., and John R. Hayes. "Cues People
      Use to Paragraph Text." Research in the Teaching of
      English 18 (1984): 147-67.
         Describes a study conducted to determine the nature
      of the "cues" that people use to paragraph texts.

563. Borland, Ron, and August Flammer. "Encoding and

Retrieval Processes in Memory for Prose." Discourse
Processes 8 (1985): 305-17.
Describes a study conducted to determine the relationship between a text's importance and the text's
stylistic features.

564. Bostian, Lloyd R. "How Active, Passive and Nominal
Styles Affect Readability of Science Writing." Journal-
ism Quarterly 60 (1983): 635-40, 670.
Examines writing styles that hinder readability.

565. Brown, Robert D., and others. "Evaluator Credibility
as a Function of Report Style: Do Jargon and Data
Make a Difference?" Evaluation Quarterly 2 (1978):
331-41.
Describes a study to determine the readability level
of reports that use jargon and data.

566. Clark, Andrew K. "Readability in Technical Writing--
Principles and Procedures." IEEE--Transactions on
Professional Communications PC-18 (1975): 67-70.
Examines the importance of readability. Discusses
three methods of determining readability levels.

567. Cleveland, William S., and Robert McGill. "Graphical
Perception and Graphical Methods for Analyzing Scientific Data." Science 229 (1985): 828-33.
Examines research conducted in the area of graphical perception. Discusses ways that viewers process
graphical information.

568. Clifford, John. "Cognitive Psychology and Writing: A
Critique." Freshman English News 13.1 (1984): 16-18.
Finds little pedagogical purpose for the information
processing model of writing.

569. David, Richard M. "A Modest Proposal." Technical
Writing Teacher 2 (1975): 1-8.
Describes an experiment currently being conducted
to "identify the effects of controlled variations in a
written technical message upon the effectiveness of the
message" (p. 3). Explains how anyone interested in
the experiment can become a participant.

570. _____. "Of Messages Sent and Received." Technical

Communication 23.4 (1976):   12-15.
Discusses the nature of messages.

571. Drury, Alinda. "Evaluating Readability." IEEE--
Transaction on Professional Communications PC-28
(1985):   11-14.
Discusses some of the problems associated with using
readability formulas.

572. Forrester, Kent. "Why Nothing Works." Teaching
English in The Two-Year College 11 (1984):   16-22.
Examines factors that contribute to the difficulty of
teaching writing.

573. Gibson, David V., and Barbara E. Mendleson. "Redundancy." The Journal of Business Communication
21.1 (1984):   43-61.
Describes researchers' attempts to study redundancy.
Examines the effect of redundancy on communication.

574. Hartley, James, and others. "Some Observations on
Producing and Measuring Readable Writing." Programmed
Learning and Educational Technology 17 (1980):   164-74.
Describes a method used to improve readability of a
sample technical writing text. Evaluates various meas-
ures that can be used to determine readability level.

575. Hines, Terence. "Left Brain, Right Brain: Who's on
First?" Training and Development Journal 39.11 (1985):
32-34.
Examines "mythology" concerning the left and right
hemispheres of the brain. Explains how research refutes the myths.

576. Hopkins, Richard L. "Educating the Right Brain: Why
We Need to Teach Patterning." Clearing House 58
(1984):   132-34.
Discusses pedological applications of findings from
research on the hemispheres of the brain.

577. Karlinsky, Stewart S., and Bruce S. Koch. "Readability Is in the Mind of the Reader." The Journal of
Business Communication 20.4 (1983):   57-69.
Describes a study to evaluate the accuracy of readability formulas.

578. Landesman, Joanne. "The 'FISAP,' Before and After."
      The Journal of Business Communication 18.4 (1981):
      17-21.
          Explains how to improve readability levels in docu-
      ments.

579. Laner, Frances J. "Readability Techniques for Authors
      and Editors." Journal of Technical Writing and Com-
      munication 6 (1976): 203-14.
          Explains how an author can work with an editor to
      produce more readable documents. States that editors
      need to know about readability research in order to
      "provide" authors with "techniques" for making docu-
      ments better suited to readers.

580. Manelis, Leon. "Effects of Prose Structure on Memory."
      Discourse Processes 6 (1983): 403-10.
          Discusses the findings of two studies that examine
      the effect of elaboration on reader retention of informa-
      tion.

581. Razek, Joseph R., and Randy E. Cone. "Readability
      of Business Communication Textbooks--An Empirical
      Study." The Journal of Business Communication 18.2
      (1981): 33-40.
          Discusses the results of a survey undertaken to
      determine the readability of business communication
      textbooks. Uses the Flesch Reading Ease Score and a
      sample of twelve textbooks to measure the writing style
      of the textbooks. Finds that "all books tested fell
      within or below the lower end of the undergraduate-
      level reading range" (p. 39).

582. "Readability Formulas: Used or Abused?" IEEE--
      Transactions on Professional Communications PC-24
      (1981): 43-54.
          Presents a series of discussions concerning the use
      of readability formulas. Evaluates various readability
      formulas.

583. Redish, Janice C., and Jack Selzer. "The Place of
      Readability Formulas in Technical Communication."
      Technical Communication 32.4 (1985): 46-51.
          Argues against the use of readability formulas.
      Suggests instead that users actually test documents.

584. Reinsch, N. L., Jr. and Phillip V. Lewis. "Communi-
      cation Apprehension as a Determinant of Channel Pref-
      erences." The Journal of Business Communication
      21.3 (1984): 53-61.
          Describes a study to determine if a communicator's
      choices of communication mode is affected by communi-
      cation apprehension.

585. Scarborough, Jule Dee. "Effects of Text Organization,
      Visuals, and Activity on Comprehension of Technical
      Discourse." Journal of Industrial Teacher Education
      22.1 (1984): 14-26.
          Describes the findings of a study of 140 junior high
      students to determine what factors affect comprehension
      and how.

586. Schenck, Eleanor M. "The Technical Writer, Reada-
      bility Formulas and the Nontechnical Reader." Journal
      of Technical Writing and Communication 7 (1977): 303-
      17.
          Discusses the importance that readability formulas
      have for the technical writer who must write for the
      nontechnical reader. Describes several readability
      studies and how information derived from these studies
      can help writers.

587. Schwartz, Helen J. "Teaching Stylistic Simplicity with
      a Computerized Readability Formula." International
      Meeting of the American Business Communication Associ-
      ation. Washington, D.C., 1980.
          Describes a study to determine how feedback con-
      cerning the readability level of their documents affected
      the style of their writing.

588. Selzer, Jack. "Certain Cohesion Elements and the
      Readability of Technical Paragraphs." Journal of Tech-
      nical Writing and Communication 12 (1982): 285-300.
          Describes a study undertaken to determine how much
      if any effect four cohesive elements--the "given/new
      contract," pronouns, synonyms and repetitions, and
      topic sentences--have on the readability of technical
      discourse.

589. _____. "Readability Is a Four-Letter Word." The
      Journal of Business Communication 18.4 (1981): 21-34.

Opposes the use of readability formulas such as the
Flesch formula and Gunning's Fog Index to "predict
the comprehensibility of a given passage" (p. 24) and
to determine rules for writing more readable documents.
Provides reasons for this opposition and discusses the
effects that readability formulas have had on teaching
practices, particularly in teaching business writing.

590. Smith, E. L., Jr. "Functional Types of Scientific
Prose." Paper presented at the Systemic Workshop.
Toronto, Canada, 25-28 August 1982.
Describes the findings of a study of scientific texts
to determine how "contextual variables" affect texts.

591. Taylor, Karl K. "The Different Summary Skills of In-
experienced and Professional Writers." Journal of
Reading 27 (1984): 691-99.
Describes a study made of professional writer, remedial college-level English students, and advanced
high school students to determine and compare the
summary skills used by each group.

592. Tibbetts, Arn. "Ten Rules for Writing Readably."
The Journal of Business Communication 18.4 (1981):
53-62.
Offers "premesis" about readability and provides ten
"rules" for making one's writing more readable. States
that the most important rule is to "Have respect for
yourself, your reader, and your language" (p. 62).

593. Vervalin, Charles H. "Checked Your Fog Index Lately?"
IEEE--Transactions on Professional Communications PC-
23 (1980): 87-88.
Explains how to use the Fog Index to determine the
readability level of one's writing.

594. Vick, Richard D. "Style in Technical Writing." Technical Writing Teacher 10 (1982): 32-34.
Discusses the importance of audience analysis to the
readability level of documents.

595. Witt, William. "Effects of Quantification in Scientific
Writing." Journal of Communication 26.1 (1976): 67-
69.
Describes a study undertaken to determine the effect

that using numerical data has on the readability level
of documents. Finds that use of numerical data in-
creases the level of difficulty of a piece of scientific
writing.

• 7. AUDIENCE ANALYSIS •

7.1 Determining the Audience

596. Campbell, David P., and Dale Level. "A Black Box Model of Communications." The Journal of Business Communication 22.3 (1985): 37-47.
   Examines various communication models and offers a new model of communication which consists of a series of processes that may or may not be known by the sender and/or receiver. Explains the implications of the model. Offers an evaluation of the model.

597. Carson, David L. "Audience in Technical Writing: The Need for Greater Realism in Identifying the Fictive Reader." Technical Writing Teacher 7 (1979): 8-11.
   Examines the complexities of audience analysis and suggests a new method for analyzing audiences.

598. Coney, Mary B. "The Implied Author in Technical Discourse." Annual Meeting of Modern Language Association. Los Angeles, California, 27-30 December 1982.
   Examines the relevance of the "implied author" concept to technical writing.

599. Dean, Morris. "Using Experimental Psychology in Technical Writing." International Technical Communication Conference. Pittsburgh, Pennsylvania, 20-23 May 1981.
   Explains how a psychological model can help writers better understand readers.

600. De Beaugrande, Robert. "Information and Grammar in Technical Writing." College Composition and Communication 28 (1977): 325-32.
   Proposes that for any piece of writing "informational

values be determined according to the presumed information level of potential readers and the efficiency
of the distribution of the new information contained in
the text" (p. 326). Suggests that teachers can better
prepare students by having them analyze texts to
determine how writers use grammatical organization to
effectively distribute information to various audiences.

601.   Dulek, Ron. "Writing to Unidentified Readers." IEEE
       --Transactions on Professional Communications PC-23
       (1980): 125-27.
           Discusses faulty assumptions that writers make when
       analyzing readers.

602.   Ewald, Helen Rothschild. "The Me in You-Attitude:
       Business Communication as Transaction." ABCA Bul-
       letin 48.1 (1985): 7-11.
           Argues that focusing mainly on readers places too
       little emphasis on writers' roles in business communica-
       tion. Explains that both writers and readers benefit
       from effective business communication; therefore, writers
       should be taught to analyze themselves as well as their
       readers.

603.   Hagge, John. "The Business Writer's Audience Is
       Rarely a Fiction." Annual meeting of the Modern
       Language Association. New York, New York, 28-30
       December 1983.
           Examines the nature of business writers' audiences.

604.   Hewes, Dean E., and others. " 'Second-Guessing':
       Message Interpretation in Social Networks." Human
       Communication Research 11 (1985): 299-334.
           Describes studies to determine how people handle
       biases when interpreting messages.

605.   Horowitz, Rosalind. "Text Patterns: Part I." Journal
       of Reading 28 (1985): 448-54.
           Examines research on readers' use of "text patterns."

606.   Leanard, Michael. "Technical Reading and Technical
       Writing." Technical Writing Teacher 10 (1983): 222-
       26.
           States that although students may be able to de-
       scribe audiences they may be unable to use this knowl-

edge to effectively produce documents for these audiences. Explains how students can learn to write more effective documents by placing themselves in positions of intended readers of the documents.

607. Levine, Leslie. "Interviewing for Information." Journal of Technical Writing and Communication 14 (1984): 55-58.
Explains how to prepare for and conduct an interview to obtain information to be used in writing technical documents.

608. Locker, Kitty O. "Theoretical Justifications for Using Reader Benefits." The Journal of Business Communication 19.3 (1982): 51-65.
Explains how "research in psychology, management, and leadership theory offers ample theoretical justification for using reader benefits" in business communications (p. 51). States that students are not aware of the theories that support the use of reader benefits and that business communication texts do not provide this theoretical background. Suggests that teachers need to present students with the theoretical background.

609. Mangan, Frank J. "The Phenomena of Technical Fallout." Journal of Technical Writing and Communication 12 (1982): 227-33.
Shows how audience needs determine the technical writer's purpose for producing a document as well as the style and format used to achieve this purpose.

610. Meyer, Bonnie J. F., and Roy O. Freedle. "Effects of Discourse Type on Recall." American Educational Research Journal 21.1 (1984): 121-43.
Describes studies to determine how various discourse types affected readers' information recall.

611. Nolan, Timothy D. "A Comparative Study of the Planning of Audience-Committed Non-Directed Technical Writing Students." Technical Writing Teacher 4 (1977): 50-54.
Describes an experiment which consisted of having students in an experimental group write technical instructions for a specified audience and students in a control group write technical instructions aimed at no

specified audience. Explains how writing for an audience increased the success of the instructions.

612.   Porterfield, Charles Donald.   "The Effects of Emotion
and Communication Skill on Message Meaning." The
Journal of Business Communication 13.3 (1976):   3-14.
Describes a study that attempts to determine "what
effects, if any, two within-receiver characteristics
(emotion and communication skill) had on the meanings
respondents gave to communicated messages" (p. 4).
Finds support for the belief that these two within-
receiver characteristics do affect the meanings of messages.

613.   Roberts, David D., and Patricia A. Sullivan.   "Beyond
the Static Audience Construct: Reading Protocols in
the Technical Writing Classroom." Journal of Technical
Writing and Communication 14 (1984):   143-53.
Suggests the use of reading protocols to teach audience awareness.

614.   Rooney, Pamela S., and Eileen B. Evans.   "Using Annual Reports to Strengthen Business Communication
Students' Understanding of Audience." ABCA Bulletin
46.4 (1983):   5-9.
Explains how annual reports are good samples of
effective audience analysis. Discusses the contents of
annual reports and defines audiences for the various
parts of annual reports.

615.   Roundy, Nancy.   "Audience Analysis: A Guide to Revision in Technical Writing." Technical Writing Teacher
10 (1983):   94-100.
Argues that "audience functions even more extensively
in revision than it does in pre-writing" (p. 94). Describes a research report project conducted to compare
the function that audience has in the writing of experienced technical writers with that in student writing.

616.   Sides, Charles H.   "Some Psychological Effects of
Syntax." Technical Communication 30.1 (1983):   14-17.
Presents the results of a study to determine how
syntax affects reader comprehension.

617.   Walzer, Arthur E.   "Articles from the 'California Divorce

Project': A Case Study of the Concept of Audience."
College Composition and Communication 36 (1985): 150-
59.

Evaluates the current heuristic for analyzing audiences. Suggests alternative.

7.2 Writing for the Audience

618. Anderson, Ulf-L. "Label Your Information." Journal
of Technical Writing and Communication 5 (1987): 295-
299.

Shows how to design articles to meet readers' needs
for information.

619. Bradford, David B. "The New Role of Technical Communicators." Technical Communication 32.1 (1985):
13-15.

Discusses changes in audiences that technical communicators must reach. Examines new strategies for
communication that will meet the changing needs of
audiences.

620. Broer, J. W. "On Two Active-Reader Stimulants:
Multiple Titles and Inverse Writing--Maximizing a Figure
of Merit for Your Publication." Journal of Technical
Writing and Communication 7 (1977): 151-157.

Uses a mathematical model to determine maximum
"merit" of a document's organization in terms of reader
response.

621. Broudy, Harry S. "Viewpoint I: On Modes of Discourse." Educational Theory 32.2 (1982): 87-88.

Discusses problems encountered by educational
philosophers as they communicate with various audiences. Suggests four levels of discourse that can be
used to solve these problems.

622. Cederbrog, Gibson A. "Write Disciplined Reports ...
That Captivate Your Readers." Chemical Engineering
82 (1975): 98-100.

Provides guidelines for writing reports to meet the
needs of readers.

623. Coney, Mary B. "The Use of the Reader in Technical

Writing." Journal of Technical Writing and Communication 8 (1978): 97-106.
    Shows the advantages of using a "mock reader" to
improve technical writing.

624.   Cosgrove, M. C. "Writing for 'The Audience.'" IEEE
     iiTransactions on Professional Communications PC-27
     1 (1984): 38-41.
       Discusses components of the writing process in
     terms of how to consider audiences in all phases of
     writing.

625.   Crodian, Bevin. "Strategies for Teaching and Adminis-
     tering Technical Writing." Journal of Technical Writing
     and Communication 10 (1980): 315-21.
       Describes the different skills needed to communicate
     with an "initiated" audience and with a lay audience.
     Explains how a misunderstanding of the skills can lead
     to ineffective technical writing programs. Finally, of-
     fers suggestions for forming more "comprehensive"
     technical writing programs that distinguish between
     these two sets of skills.

626.   De Beaugrande, Robert A. "Communication in Techni-
     cal Writing." Journal of Technical Writing and Com-
     munication 8 (1978): 5-15.
       Discusses what constitutes effective communication
     in terms of determing what constitutes "new" informa-
     tion and "old" information for readers.

627.   Dulek, Ron. "Writing to Unidentified Readers." IEEE
     --Transactions on Professional Communications PC-23
     (1980): 125-127.
       States that when writing to unidentified readers,
     writers must make assumptions. Cautions writer not
     to make faulty assumptions about the unidentified
     reader and lists four faulty assumptions writers fre-
     quently make. Provides strategies to counteract these
     faulty assumptions.

628.   Dunkle, Susan B., and Purvis M. Jackson. "Informa-
     tion Structuring: Relating Old and New Knowledge."
     IEEE--Transactions on Professional Communications
     PC-25 (1982): 175-77.
       Explains how information structuring can be used to
     present technical concepts to nontechnical readers.

629. Fuller, Don. "How to Write Reports That Won't Be
      Ignored." IEEE--Transactions on Professional Com-
      munications PC-23 (1980): 79-81.
         States that one usually writes to either inform or
      influence. Suggests that when writing to inform, the
      writer should make meanings clear, present accurate
      data and show significance of the information being
      presented. Further suggests that when writing to in-
      fluence, the writer should "tailor" writing approach to
      audience's attitude toward the subject, determine the
      slope of the report, and give emphasis to the important
      items in the report.

630. Herrstrom, David Sten. "A Matrix of Audience Re-
      sponses for the Internal Proposal." Technical Writing
      Teacher 10 (1983): 101-7.
         Finds that most textbooks discuss proposals "written
      not from within an organization but from outside" (p.
      101). Describes the relationships that exist between
      writers and readers of internal proposals. Offers
      strategies that proposal writers can use to influence
      audience responses to internal proposals.

631. Johnson, Parker H.; Judith Carrard; and William Haus-
      man. "A Method for Translating Technical to Non-
      technical Prose." Journal of Technical Writing and
      Communication 12 (1982): 191-200.
         Outlines and describes a five-step method that
      writers can use to translate technical language into
      prose that lay audiences can understand. Also, offers
      a way to test the "validity" of the nontechnical prose.

632. Jordan, Michael P. "Credibility and Reasoning in Tech-
      nical Writing--Some Notes for Writers, Editors, and
      Instructors." Journal of Technical Writing and Com-
      munication 6 (1976): 35-51.
         Discusses the need for understanding the role that
      credibility and logic play in the writing of technical
      reports. States that the best way to teach students
      how to handle credibility and logic in their writing is
      to have students analyze actual documents. By examin-
      ing actual documents students will see how the need
      for credibility and the use of logic depends on the
      nature of the reader and the purpose of the document.
      Thus, students learn when they need to add more
      credibility to their writing and how to do so.

633. Kerfoot, Glenn, W. "Tips on Technical Writing."
      IEEE--Transactions on Professional Communications
      PC-20 (1977): 17-19.
          States that technical writers must keep their readers
      in mind. Suggests that writers can improve their writ-
      ing by 1) using "simple" terms, 2) knowing rules of
      capitalization, 3) using active verbs, 4) using concrete
      terms, and 5) having others proofread the document.

634. Klausen, James. "The Mulholland Test for Understanda-
      bility in Popular Science Writing." English Journal
      67.4 (1978): 72-3.
          Discusses problems that writers must address when
      presenting scientific topics to the average reader. Ex-
      plains how writers can transform sophisticated vocabu-
      lary and complex ideas into words and concepts that
      average readers easily understand. Warns that simpli-
      fication should not lead to misrepresentation of informa-
      tion.

635. Kogen, Myra. "The Role of Audience in Business and
      Technical Writing." ABCA Bulletin 46.4 (1983): 2-4.
          Discusses ways to incorporate information learned
      from studies of audience analysis into the business and
      technical writing classroom.

636. Lazaro, Timothy R. "Effective Communication of Tech-
      nical Information to a Nontechnical Group." Journal of
      Technical Writing and Communication 7 (1977): 295-
      301.
          Describes a five-step method for preparing technical
      information for presentation to a nontechnical group.
      To prepare such a presentation, the writer/presenter
      must 1) determine a rationale, 2) select goals, 3) con-
      sider the political situation, 4) develop an argument
      and determine recommendations based on the political
      situation and attitudes of the audience, and 5) prepare
      the presentation. Gives specific examples to show how
      to use this five-step method to prepare a presentation.

637. Lufkin, James. "The Gulf Between Correctness and
      Understanding." IEEE--Transactions on Professional
      Communications PC-19 (1976): 4-6.
          Shows how audience analysis suffers when too much
      emphasis is placed on "correctness" and "clarity."

638. Mann, James A., and John B. Ketchum. "Multi-Use:
      Reshaping Technical Material." IEEE--Transactions on
      Professional Communications PC-27 2 (1984):   85-8.
          Explains how to take existing communication material
      and change the material according to varying audience
      needs.

639. Ottilio, C. J. "A Scientific Releaser." Journal of
      Technical Writing and Communication 6 (1976):   259-61.
          Discusses the importance of informing laymen of
      scientific advancements. Discusses various methods of
      technical communication that can be used to present
      scientific information to lay audiences.

640. Paul, Terri, and Mary Rosner. "Discovering and Teach-
      ing Syntactic Structures in Three Technical Disciplines."
      Journal of Technical Writing and Communication 13
      (1983):   109-22.
          Presents an analysis of technical writing for three
      different audience levels to determine if different tech-
      nical styles exist.

641. Powell, Kenneth B. "Writing with Verve." IEEE--
      Transactions on Professional Communications PC-25
      (1982):   189-191.
          Describes how to write for the lay audience.

642. Raudsepp, Eugene. "Present Your Ideas Effectively."
      IEEE--Transactions on Professional Communications
      PC-22 (1979):   204-10.
          Explains how the method of presenting ideas affects
      readers'/listeners' acceptance of the ideas. Provides
      guidelines for presentation of ideas.

643. Russo, Angelo Del. "Sensible Writing Can Be Under-
      stood." Journal of Technical Writing and Communica-
      tion 9 (1979):   169-72.
          Discusses the importance of audience analysis in
      technical writing. Provides specific examples to help
      explain why technical writers must write so that the
      reader can understand the material. Also, suggests
      how technical writers can make their works more read-
      able (understandable). Emphasizes the fact that the
      writer must thoroughly understand the subject matter
      so that he can present that subject matter to different
      audiences.

644. Saunders, Arlan. "Writing Effective Assembly Proce-
      dures." IEEE--Transactions on Professional Communi-
      cations PC-25 (1982):   20-21.
          Discusses the importance of meeting readers' needs
      when writing procedures.

645. Schenck, Eleanor M. "The Technical Writer--Readability
      Formulas and the Nontechnical Reader." Journal of
      Technical Writing and Communication 7 (1977):   303-317.
          Discusses strategies for presenting information to
      nontechnical readers. Examines the use of readability
      formulas for evaluating documents.

646. Soldow, Gary F. "A Study of the Linguistic Dimension
      of Information Processing as a Function of Cognitive
      Complexity." The Journal of Business Communication
      19.1 (1982):   55-69.
          Describes a study "concerned with the importance
      of tailoring the syntactical receiver in light" of the
      reader's level of ability to process "syntactically com-
      plex sentences" (55-56). Results of the study show
      that the ability to "tailor" syntactical structure of a
      message to meet a reader's "cognitive complexity" is
      important for writers trying to produce effective com-
      munication.

647. Tagliacozzo, Renata. "Some Stylistic Variations in
      Scientific Writing." Journal of the American Society
      for Information Science 29 (1978):   136-40.
          Examines two groups of scientific articles that focus
      on the same subject and determines what stylistic varia-
      tions exist when writing at different levels of special-
      ization.

648. Whitburn, Merrill D. "Personality in Scientific and
      Technical Writing." Journal of Technical Writing and
      Communication 6 (1976):   299-306.
          Examines the importance of creating "personality"
      in scientific and technical writing. Explains how such
      writing reflects the growing importance of the "human
      element" in science. Suggests that technical writing
      teachers can use scientific writing from the past to
      teach students how "personality" can effectively be
      used in scientific and technical writing.

649. Whittaker, Della A. "Write for Your Reader." IEEE--
Transactions on Professional Communications PC-23
(1980): 170-73.
  Explains how to write to meet readers' needs.

650. Williams, Robert I. "Playing with Format, Style, and
Reader Assumptions." Technical Communication 30.3
(1983): 11-13.
  Views technical writing as a "game " played between
writer and reader.

• 8. STYLE •

8.1 General Guidelines

651. Annett, Clarence H. "An Editor's View: Ten Common Errors in Technical Writing." Journal of Technical Writing and Communication 12 (1982): 185-90.
Lists and disucsses ten errors that occur in technical writing.

652. Battison, Robbin, and Dixie Goswami. "Clear Writing Today." The Journal of Business Communication 18.4 (1981): 5-16.
Examines the problems that have contributed to unclear writing. Determines what constitutes clear writing and suggests how clear writing can be achieved. Offers suggested teaching methods to help writers produce clear writing.

653. Bradner, Lowell; Orville Bidwell; and Iwan Teare. "Let Logic Guide Your Writing." Journal of Technical Writing and Communication 10 (1980): 347-54.
Provides suggestions for producing clear, concise, writing.

654. Broer, Jan W. "Linking While Writing--Do You or Don't You?" Journal of Technical Writing and Communication 8 (1978): 217-225.
Examines the causes of incoherent texts and suggests how to create coherent texts.

655. Brogan, Katherine M., and James D. Brogan. "Aspects of Good Writing: An Engineer's Perspective." CEA Forum 13.2 (1982): 5-9.
Examines what constitutes good technical writing.

656. Carroll, John E. "Technical Writes and Wrongs." IEEE
      --Transactions on Professional Communications PC-23
      (1980): 68-71.
         Discusses guidelines for producing good technical
      writing.

657. Donnellan, LaRai M. "Technical Writing Style: Pref-
      erences of Scientists, Editors and Students." Confer-
      ence on CCC. Minneapolis, Minnesota, 21-23 March
      1985.
         Describes a study conducted to determine writing
      style preferences.

658. Eisenberg, Anne. "Technical and Readable." Gradu-
      ate Engineer 7.2 (1985): 51-53.
         Suggest ways that engineers can improve their writ-
      ing style.

659. Estrin, Herman. "How to Write for Scientific and Tech-
      nical Journals." The Journal of Business Communica-
      tion 18.3 (1981): 55-58.
         Offers guidelines for choosing subjects and using
      writing styles appropriate for publication in scientific
      and technical journals.

660. Farkas, David K. "The Concept of Consistency in
      Writing and Editing." Journal of Technical Writing and
      Communication 15.4 (1985): 353-64.
         Defines what is meant by "consistency" and discusses
      the role of consistency in technical communication.

661. Feineman, George. "How to Write Like a Logistician
      Should." Transactions on Professional Communication
      PC-22 (1979): 200-01.
         Discusses common hindrances to communication and
      explains how to produce effective communication.

662. Franxman, James J. "Technical Writing--Preparing
      Scientists for a Career in Industry." Technical Writing
      Teacher 10.3 (1983): 186-88.
         Discusses "some of the writing problems" that the
      author has seen "in the reports of young (and some-
      times not so young) scientists" (p. 186). Covers the
      following topics: 1) faulty audience analysis, 2) build-
      ing suspense in reports, 3) including every experiment

in reports, 4) mixing results and discussion, and 5)
spelling words incorrectly.

663. Giermanski, James R. "Writing Writing for Writing's
      Sake." Technical Writing Teacher 9 (1981): 31-34.
          Examines five style manuals to show how varied
      "rules" of form are for "literary and scientific disci-
      plines with respect to the submission of research and
      other papers" (p. 31). Argues that content considera-
      tions should be the most important criteria for judging
      the acceptability of an article for publication.

664. Gleason, James P. "Humor Can Improve Your Technical
      Presentations." IEEE--Transactions on Professional
      Communications PC-25 (1982): 86-90.
          Shows how humor can be used effectively in techni-
      cal publications.

665. Hogan, J. B. "Statistical Doublespeak: The Deceptive
      Language of Numbers." Technical Writing Teacher 10
      (1983): 126-29.
          Describes how statistics can be manipulated to de-
      ceive readers. Provides examples that show the "mis-
      use" of statistics. Suggests guidelines to help writers
      "determine the accuracy and appropriateness of any
      statistics" (p. 127). States that technical writing
      teachers help students learn how to correctly use
      statistics in their writings.

666. Hulbert, Jack E. "Communication: Effective Business
      Writing." Business Education Forum 34.5 (1980): 20,
      22-24.
          Provides guidelines for producing effective business
      writing.

667. Hurley, Michael C. "The Importance of Style in Writ-
      ing." Media and Methods 16.6 (1980): 34-36.
          Explains that "style" consists of more than gram-
      matical correctness, simplicity, and clarity.

668. Jordan, Michael P. "As a Matter of Fact." The Jour-
      nal of Business Communication 15.2 (1978): 3-12.
          Discusses how writers determine what data to con-
      sider fact and what data to consider opinion.

669. Lippincott, H. F. "Some Tips for Clear Writing." IEEE
     --Transactions on Professional Communications PC-26
     (1983): 11-14.
         Provides guidelines for producing good writing.

670. Marder, Daniel, and Dorothy Margaret Guinn. "De-
     fensive Aesthetics for the Technical Writer." Journal
     of Technical Writing and Communication 12 (1982): 35-
     42.
         Examines the issue of technical accuracy versus ef-
     fective style. Suggests ways of achieving both.

671. McDonald, Daniel. "Don't Go Like That: A Guide to
     Correct Writing." Journal of Technical Writing and
     Communication 8 (1978): 227-31.
         Offers guidelines for producing clear, concise writ-
     ing.

672. Messer, Donald K. "Six Common Causes of Ambiguity."
     Technical Writing Teacher 7 (1980): 50-52.
         Lists six types of ambiguity that often are found in
     technical prose. Suggests ways to eliminate ambiguity
     in writing.

673. Mueller, Douglas. "Put Clarity in Your Writing." IEEE
     --Transactions on Professional Communications PC-23
     (1980): 173-78.
         Explains how to produce clear writing.

674. Odell, Lee. "Business Writing: Observations and Im-
     plications for Teaching Composition." Theory into
     Practice 19.3 (1980): 225-232.
         Examines the state of current writing in the busi-
     ness world by presenting samples of poor writing.
     Shows how these writing samples can be improved.

675. Plung, Daniel L. "Add Style to Your Technical Writ-
     ing." IEEE--Transactions on Professional Communica-
     tions PC-27 1 (1984): 20-24.
         Suggests some figurative devices that can be used
     to improve technical writing style.

676. _____. "Literacy Communications: Adding Style to
     Professional Writing." Journal of Technical Writing and
     Communication 10.1 (1982): 25-31.

Discusses literacy devices that can be appropriately used in technical writing and provides suggestions for how to use these devices effectively.

677.   Ramsey, Richard David.   "Technical Writing Stylistics, and TG Grammar."   Journal of Technical Writing and Communication 7 (1977):   333-345.
Explains how transformational generative grammar can be used to teach technical writers "to understand and avoid the pitfalls in an impersonal style" (p. 344).

678.   Rathjens, Dietrich.   "The Seven Components of Clarity in Technical Writing."   IEEE--Transactions on Professional Communications PC-28 (1985):   42-46.
Discusses what constitutes clarity in technical writing.

679.   Rodman, Lilita.   "Strategies for Removing Ambiguity in Technical Writing."   Conference on CCC.   Minneapolis, Minnesota, 5-7 April 1979.
Describes a method for analyzing ambiguous writing in order to determine how to eliminate ambiguity from writing.

680.   Rook, Fern.   "Syntax Error."   Technical Communication 31.3 (1984):   38.
Explains how to be more specific when defining errors found in technical writing.   States that being more specific makes correcting the errors more easy.

681.   Sawyer, Thomas M.   "Explanation."   Journal of Technical Writing and Communication 15 (1985):   131-41.
Suggests strategies that technical writers can use to explain difficult scientific and engineering topics.

682.   Sims, Diana Mae.   "What Is Technical English, American Style?"   Combined Annual Meetings of the American Dialect Society and New Ways of Analyzing Variation. Washington, D.C., 2-4 November 1978.
Discusses the development of American scientific/ technical style.   Describes the characteristics of this style.

683.   Sloan, Gary.   "An English Teacher Takes on the Technicians."   Technical Writing Teacher 5 (1978):   50-52.

Suggests ways to eliminate common stylistic problems that often plague technical writing in order to produce clear writing.

684. Stratton, Charles R. "Analyzing Technical Style." Technical Communication 26.3 (1979): 4-8.
    Analyzes numerous technical writing samples to determine what constitutes good style in technical writing.

685. Swain, Philip W. "Giving Power to Words." IEEE-- Transactions on Professional Communications PC-23 (1980): 135-137.
    States that audience awareness affects word choice. Provides rules for making writing simple, direct and concise.

686. Tagliacozzo, Renata. "Levels of Technicality in Scientific Communication Information." Processing and Management 12.2 (1976): 95-110.
    Discusses a comparison made between science writing found in Scientific American and writing found in the scientific articles cited as references in Scientific American. Points out similarities and differences in these two types of science writings.

687. Tan San Yee, Christine. "Sequence Signals in Technical English." RELC Journal 6.2 (1975): 63-101.
    Explains that coherence is achieved in technical English by the use of "sequence signals." States that students should be taught to recognize these "linguistic forms" in order to understand how coherence is achieved in technical writing.

688. Tibbetts, Arn. "What Are the Most Useful Principles for Teaching Business Writing." ABCA Bulletin 47.3 (1984): 18-21.
    Presents ten "principles" for producing effective written communication.

689. Timm, Paul R., and Daniel Oswald. "Plain English Laws: Symbolic or Real?" The Journal of Business Communication 22.2 (1985): 31-38.
    Questions the "effectiveness of legislation in producing understandable documents" (p. 33). Conducts a survey of 846 members of the Association for Business

Communication. Results of that survey, which are dis-
cussed in the article, lead the authors to conclude that
plain English laws are not effective. Describes three
problems that must be addressed if the laws are to be
effective. Offers suggestions for how to "overcome"
these problems.

690. Velte, Charles. "The Treatment of Technical Names."
IEEE--Transactions on Professional Communications
PC-20 (1977): 23-26.
Provides guidelines for handling capitalization,
underscoring and italics, jargon and other conventions
used in technical writing. Shows why overuse of such
conventions hinders clarity of technical writing.

691. Whittaker, Della A. "Remember the Tradition?" Journal of Technical Writing and Communication 7 (1977):
191-95.
Reminds writers of some traditional rules of style
that promote clarity in writing.

692. Wight, Eleanor. "How Creativity Turns Facts into
Usable Information." Technical Communication 32.1
(1985): 9-12.
Examines the role of creativity in technical communication. Shows how creative strategies can enhance the
effect of communication.

693. Wilkins, Keith A. "Technical Writing: Effective Communication." Journal of Technical Writing and Communication 7 (1977): 35-43.
Describes what constitutes good technical writing.

694. Woolsey, Gene. "Corporate Style, Corporate Substance,
and the Sting." IEEE--Transactions on Professional
Communications PC-23 (1980): 67-68.
Details a story to show the importance of good writing.

695. Zeidner, Martin A. "Physician, Heal Thyself." Journal
of Technical Writing and Communication 9.3 (1982):
186-87.
Discusses characteristics that make for ineffective
technical writing.

8.2 Rhetorical Considerations

696. Bernhardt, Stephen A. "The Writer, the Reader, and
Scientific Text." Journal of Technical Writing and
Communication 15 (1985): 163-74.
Examines the rhetorical elements of scientific dis-
course. Discusses seven writing situations which ex-
hibit "personal intrusions" by the writers.

697. Brown, Byron. "Rhetorical Figures of Communion,
Presence, and Severance: Toward an Efficient Rhetoric of Business." ABCA Bulletin 42.3 (1979): 14-17.
Shows how combining concepts from current rhetorical theory with the figures of speech from classical
rhetoric can produce effective writing.

698. Bump, Jerome. "Metaphor, Creativity, and Technical
Writing." College Composition and Communication 36
(1985): 444-53.
Points out that many technical writing textbooks
omit discussions of metaphor, thus leading students to
think that the use of metaphor is not appropriate in
technical writing. Shows through the citing of numerous examples, how scientist have used metaphor to
make their writing more exact and easier to comprehend.

699. Corey, Robert L. "Rhetoric and Technical Writing:
Black Magic or Science?" Technical Communication
25.4 (1978): 2-6.
Examines the role of rhetoric in technical writing.

700. De Bakey, Lois, and Selma De Bakey. "The Art of
Persuasion: Logic and Language in Proposal Writing."
Grants Magazine 1.1 (1978): 43-60.
Examines the steps involved in preparing an effec-
tive proposal, beginning with selection of an agency
and ending with producing the actual written proposal.

701. Girill, T. R. "Technical Communication and Rhetoric."
Technical Communication 32.1 (1985): 44.
Examines the relationship between rhetoric and tech-
nical communication.

702. Gooch, James L. "Knowing, Communicating, and Scientific Rhetoric." Biological Science 25.11 (1975): 715-
17.

States that scientific rhetoric be used in the writing
of scientific documents.

703. Grimshaw, James A., Jr. and William E. McCarron.
     "Hidden Persuasions in Technical Writing." Technical
     Writing Teacher 9 (1981):   19-22.
         Argues that "Even the blandest, most seemingly
     objective technical writing has a persuasive slant either
     by its emphasis, its omission of other possible alterna-
     tives, or its use of scientific language" (p. 21). Sup-
     ports this argument by showing how ethos, pathos,
     and logos are used in rhetorical situations that involve
     the use of technical writing.

704. Guinn, Dorothy Margaret.   "Ethos in Technical Dis-
     course." Technical Writing Teacher 11 (1983):   31-37.
         Argues that although they try, technical writers
     cannot eliminate "individual personality" from their
     writing.   Contends that stylistic choices assist in giving
     "personality" to writing and that this "individuality"
     can be beneficial.

705. Halloran, S. Michael.   "Technical Writing and the Rhe-
     toric of Science." Journal of Technical Writing and
     Communication 8 (1978):   77-88.
         Examines the traditional view that rhetoric and sci-
     ence are separate areas of concern.   Uses Watson's
     The Double Helix to show that science is rhetorical.

706. Harris, John Sterling.   "Metaphor in Technical Writing."
     Technical Writing Teacher 2 (1975):   9-13.
         States that although use of metaphor in scientific
     and technical writing is frowned upon, actual practice
     shows that metaphor is commonly used, and used ef-
     fectively, in such writing.   Provides general guide-
     lines for the use of the "technical metaphor."

707. Kallendorf, Craig, and Carol Kallendorf.   "The Figures
     of Speech, Ethos, and Aristotle: Notes Toward a Rhe-
     toric of Business Communication." The Journal of
     Business Communication 22.1 (1985):   35-50.
         Shows that practicing business writers effectively
     employ classical figures of speech in their business
     writing.   Details a "fourth kind of rhetoric for business"
     (p. 45) that is an expansion of Aristotelian deliberative,
     judicial, and epideictic rhetorics.

708. Limaye, Mohan R. "Improving Technical and Bureau-
      cratic Writing." Journal of Technical Writing and Com-
      munication 11 (1981): 23-33.
      Discusses "four syntactic-rhetorical principles which
      make written discourses easier to read or faster to
      process" (p. 23): 1) keep subjects and verbs together;
      2) use appropriate prepositions between nouns to indi-
      cate the semantic relationship between them; 3) help
      readers to segment syntactic units correctly; and 4)
      match textual sequence with chronological sequence.

709. Marder, Daniel. "Technical Reporting Is Technical
      Rhetoric." Technical Communication 25.4 (1978): 11-
      13.
      Explains how technical writers must be able to ef-
      fectively use rhetoric to create persuasive documents.

710. Minor, Dennis E. "A Concept of Language for Techni-
      cal Writing Students." Journal of Technical Writing
      and Communication 8.1 (1978): 43-51.
      States that technical writing students need to under-
      stand not only how to make their language grammati-
      cally correct but also how to make their language mean-
      ingful. Describes three "functions" of language in
      technical writing: generative, substantive, and con-
      clusive. Suggests how teachers can help technical
      writing students learn how to use language to achieve
      these functions.

711. Plung, Daniel L. "The Motivated Sequence and the
      Environmental Impact Statement." Journal of Technical
      Writing and Communication 10 (1980): 65-72.
      Discusses a writing strategy termed the "motivated
      sequence." Explains how this strategy simulates human
      thought process to produce more effective written com-
      munication. Shows how use of the motivated sequence
      when writing Environmental Impact Statements produces
      better writing than adherence to guidelines for writing
      that the National Environmental Policy Act established.

712. Sachs, Harley L. "Rhetoric, Persuasion, and the Tech-
      nical Communicator." Technical Communication 25.4
      (1978): 14-15.
      Discusses attitudes toward the use of rhetoric in
      technical writing.

713. Shimberg, H. Lee. "Ethics and Rhetoric in Technical
      Writing." Technical Communication 25.4 (1978): 16-18.
          Discusses deceptive uses of rhetoric. States that
      the technical writer is responsible for correctly using
      rhetorical strategies.

714. Stevenson, Dwight W. "Toward a Rhetoric of Scien-
      tific and Technical Discourse." Technical Writing
      Teacher 5 (1977): 4-10.
          Discusses what work has been done in the field of
      "technology assessment." States that "work in rhetoric
      --invention and arrangement--needs to be done" in
      regard to this field (p. 17). Suggests that studies
      should be undertaken to develop a rhetoric of "use of
      visuals in scientific and technical communication" as
      well as a rhetoric of technical manuals (pp. 7-8).

8.3 Grammar and Punctuation

715. Bernstein, Theodore M. "Punctuation." IEEE--Trans-
      actions on Professional Communications PC-20 (1977):
      38-44.
          Lists and discusses rules for correct punctuation.

716. Bush, Don. "The Case of the Inconsistent Upper Case
      (A Model Solution)." Technical Communication 31.1
      (1984): 11-13.
          Suggests some guidelines for how to correctly use
      capital letters.

717. Farkas, David K. "The Use of Quotation Marks and
      Italics to Introduce Unfamiliar Terms." Journal of
      Technical Writing and Communication 13 (1983): 369-
      74.
          Explains why and how writers use quotation marks
      and italics to mark words thought to be unfamiliar to
      readers.

718. Foley, Louis. "Plain Comma Sense." Journal of Tech-
      nical Writing and Communication 5 (1975): 287-94.
          Describes an approach for determining when and how
      to use punctuation.

719. Limaye, Mohan R. "Approaching Punctuation as a

System." ABCA Bulletin 46.1 (1983): 28-33.
Examines common rules of punctuation and discusses
two "principles" that are often overlooked. Explains
how students can be taught to understand "the close,
logical relationship between syntax and punctuation"
(p. 32).

720. McDonald, Daniel. "'Don't Go Like That': A Guide to
Correct Writing." Journal of Technical Writing and
Communication 8 (1978): 227-31.
Provides a system to help writers avoid making com-
mon grammatical errors in their writing.

721. Montoya, Sarah. "Word Watching: Who or Whom?"
IEEE--Transactions on Professional Communications
PC-24 (1981): 84-85.
Describes an exercise to help writers decide when
to use "who" or "whom."

722. Orth, Mel F. "Striking Out: Poor Style and Grammar
Still Abound in Technical Writing." IEEE--Transactions
on Professional Communications PC-21 (1978): 44-47.
Discusses problems with style and grammar that are
found in technical writing.

723. Ramsey Richard David. "Improving Literacy Through
Memorizing Patterns." Technical Writing Teacher 10
(1983): 218-21.
Presents step-by-step instructions that explain how
technical writing teachers can use quizzes to "improve
... use of syntax, punctuation, and spelling" (p. 218)
in their students' writing. Includes a sample quiz and
answer key.

724. Rook, Fern. "Commonplace Commas." Technical Communication 31.2 (1984): 29.
Examines the various uses of commas in writing.

725. _____. "Just in Case." Technical Communication
32.3 (1985): 45.
Examines pronoun usage. Provides samples of correct and incorrect uses of pronouns.

726. _____. "Piled-up Modifers." Technical Communication 31.4 (1984): 63.

Discusses correct and incorrect usage of modifiers in writing.

727. _____. "Who Cares About Whom?" Technical Communication 32.4 (1985): 67-68.
Discusses the use of "who" and "whom" in writing.

728. Tibbetts, A. M. "Do Your Students Need 'Grammar'?" The Journal of Business Communication 12.4 (1975): 3-9.
Examines the role of grammar in effective writing. Suggests that grammar does matter in that it should be used as the "support" of good writing. Gives suggestions for incorporating grammar instruction into writing courses.

729. Waltman, John L. "Grammar Competency and Business Communications." Meeting of the Southwest Division of the American Business Communication Association. Houston, Texas, 10-12 March 1983.
Describes a study conducted to determine how grammar competency affects job success.

8.4 Diction

730. Billingsley, Patricia A., and Neil A. Johnson. "Nonsexist Use of Language in Scientific and Technical Writing." IEEE--Transactions on Professional Communications PC-22 (1979): 193-98.
Offers guidelines to help writers eliminate sexist language from their documents.

731. Bohlman, Herbert M., and Alan P. Wunsch. "Common Sense Writing: A Developing Trend." Journal of Business Education 57.3 (1981): 95-97.
Examines the trend to reduce jargonistic, poorly structured writing.

732. Briden, Earl F. "The Jargonist as Comedian: An Approach to Usage." ABCA Bulletin 45.1 (1982): 39-41.
Describes problems that arise from use of jargon and convoluted syntax. Suggests ways to eliminate these problems.

733. Brown, William R. "Jargon and the Teaching of Organ-
     izational Communication." American Business Communi-
     cation Association Eastern Regional Meeting. Philadel-
     phia, Pennsylvania, 21-22 April 1983.
         Examines the importance of jargon in technical com-
     munication.

734. Butenhoff, Carla. "Bad Writing Can Be Good Business."
     ABCA Bulletin 40.2 (1977): 12-13.
         Argues that sometimes business writing needs to be
     vague. Discusses how to write effective memos.

735. Cline, Carolyn Garrett, and Lynn Masel-Walters. "At
     Least the Editors Are TRYING: Women and Sexism in
     Corporate Publications." ABCA Bulletin 46.3 (1983):
     26-30.
         Presents the results of a survey of twenty-nine in-
     house publications to analyze the use of sexist language
     in the publications.

736. Damerst, William A. "Students Aiming for a Larger
     Audience May Close 'All' Language Gaps." ABCA Bul-
     letin 45.1 (1982): 5-7.
         Discusses what teachers can do toward reaching a
     compromise between jargonistic writing and plain Eng-
     lish.

737. Draycott, David. "Commercialese--A Form of Words?"
     English in Education 14.2 (1980): 15-17.
         Explains the importance of teaching students how to
     appropriately use "Commercialese" when writing busi-
     ness communications.

738. Fischborn, Herb. "In Defense of the Language."
     IEEE--Transactions on Professional Communications
     PC-27 (1984): 7-9.
         Provides examples to show how words can be mis-
     used.

739. Foley, Louis. "Malapropism Becomes Common Place."
     Journal of Technical Writing and Communication 6 (1976):
     129-33.
         Discusses the origin and use of malapropisms in
     numerous kinds of writing.

740. Gallagher, Brian. "Vocabulary in Writing for Business:
      Six Propositions for Pedagogical Use." Journal of Basic
      Writing 2.3 (1979):   40-58.
          Discusses six "propositions" for determining vocabu-
      lary to be used in business communication. Includes
      sample assignments.

741. Gilsdorf, Jeanette W. "Executive and Managerial At-
      titudes Toward Business Slang: A Fortune-List Sur-
      vey." The Journal of Business Communication 20.4
      (1983):   29-42.
          Describes a survey of communicators in top-ranked
      organizations that examines their attitudes concerning
      the use of jargon.

742. Goldman, Louis. "The Logic of Euphemisms and the
      Language of Education." Educational Theory 26 (1976):
      182-87.
          Explains when and how euphemisms can be used ef-
      fectively in writing. Also, examines instances when
      the use of euphemisms is ineffective.

743. Harris, John S. "The Naming of Parts: An Examina-
      tion of the Origins of Technical and Scientific Vocabu-
      lary." Journal of Technical Writing and Communication
      14 (1984):   183-91.
          Discusses various sources of technical and scientific
      vocabulary. Explains when use of this vocabulary is
      appropriate in writing.

744. Hays, Robert. "The Trade Jargon of Proposal Writing:
      A Brief Glossary." Technical Writing Teacher 11 (1984):
      94-99.
          Discusses the importance of teachers' knowing terms
      used in proposal writing. Provides a selective glossary
      of these terms, focusing on "working definitions, not
      legal refinements" (p. 95).

745. McDonald, Daniel. "Be a Better Writer (Don't Use
      Dirty Words)." Journal of Technical Writing and Com-
      munication 7 (1977):   183-90.
          Shows how poor word choices produce poor writing.

746. _____. "The Fifty Best Words." Journal of Tech-
      nical Writing and Communication 14 (1984):   351-56.

Presents a list of the "best" words that can be used by writers.

747. Meyers, Manny. "The Founding Fathers: A Grand Case of Semantic Distortion." Journal of Technical Writing and Communication 6 (1976): 53-55.
Explains why the phrase " 'Founding Fathers' is a prime case of how a phrase can undergo radical alteration not by evolution but by purposeful distortion" (p. 53).

748. Pickrel, Paul. "Identifying Clichés." College English 47 (1985): 252-61.
Describes what cliches are and how they develop.

749. Saliskas, Joan. "The Man-Machine Interface: Verbiage in the Computer Age." ABCA Bulletin 47.4 (1984): 12.
Suggests that writers must learn to eliminate "verbiage" from their writings, which in turn will produce computer languages that are more user friendly.

750. Tibbetts, Charlene. "Sex and Language." ABCA Bulletin 46.3 (1983): 24-26.
Examines the issue of sexism in language. Suggests guidelines for using the generic pronouns "he" and "his."

751. Velte, Charles. "Treatment of Technical Names." Technical Communication 22.3 (1975): 6-8.
Discusses problems that result from use of technical terms. Provides guidelines for using technical terms in writing.

8.5 Sentence Structure

752. Broadhead, Glenn J. "Sentence Skills for Technical Writers." Journal of Technical Writing and Communication 11 (1981): 139-50.
Examines technical writers' actual practices concerning the formation of sentences.

753. Chard, John. "The Incidence of Sentence Openers in Selected Technical or Scientific Periodicals and Journals." Journal of Technical Writing and Communication

5 (1975): 91-98.

Describes a study which uses Christensen's categories of "sentence openers" to determine what openers
published writers used for their sentences. Explains
how information obtained from the study can be used
by student writers.

754. Conway, William D. "The Passive Voice: Friend of
Foe of the Technical Communicator?" Technical Writing
Teacher 8 (1981): 86-90.

States that technical writing teachers need to explain to students the difference between active and passive voice and help students learn to use "both the
active and passive effectively (p.86). Examines
problems that can occur from the use of passive voice.
Offers guidelines for determining when to use active
or passive voice.

755. Douglas, George H. "What to Do About Cobblestone
Writing." Technical Writing Teacher 5 (1977): 18-21.

Defines "cobblestone writing" as that which "suffers
from hardness and density" (p. 18). Gives samples of
such writing and suggests ways to eliminate it.

756. Johnson, Thomas P. "How Well Do You Inform?" IEEE
--Transactions on Professional Communications PC-25
(1982): 5-9.

Discusses the importance of using details to support
general statements. Suggests ways to improve sentence
structure.

757. Kies, Daniel. "Some Stylistic Features of Business and
Technical Writing: The Functions of Passive Voice,
Nominalization, and Agency." Journal of Technical
Writing and Communication 15 (1985): 299-308.

Discusses relationships between passive voice, nominalization, and agency in technical prose style. States
that writers need to understand this relationship.

758. Ramsey, Richard David. "Grammatical Voice and Person in Technical Writing: Results of a Survey." Journal of Technical Writing and Communication 10 (1980):
109-13.

Describes a survey conducted to answer the question:
"Is the third person passive superior for communication

of scientific fact?" (p. 110). Finds that use of the third person passive is "disadvantageous from a communication standpoint" (p. 112).

759. Reimold, Cheryl. "Business Writing--Clear and Simple." IEEE--Transactions on Professional Communications PC-24 (1981): 184-85.
Suggests ways to eliminate wordiness.

760. Rodman, Lilita. "The Passive in Technical and Scientific Writing." Conference on CCC. Dallas, Texas, 26-28 March 1981.
Discusses a study of journals to determine how the passive voice was used in writing.

761. Schoenfeld, Robert. "When to Make Nouns Go into Action." IEEE--Transactions on Professional Communications PC-24 (1981): 123.
Suggests converting nouns to verbs to produce better writing.

762. Sears, Jeffry S. "Jargon and Gobbledygook: A Checklist of Symptoms." ABCA Bulletin 42.3 (1979): 25-28.
Identifies problems with jargon and awkward sentence structure that occur in prose.

763. Smith, Frank R. "Verbal Pollution." IEEE--Transactions on Professional Communications PC-22 (1979): 165-69.
Discusses four types of "verbal pollution." Presents causes of and remedies for wordiness.

764. Stewart, Roy. "Writers Overuse the Passive Voice." Technical Communication 31.1 (1984): 14-16.
Shows that "writers of electronic-instrument technical manuals" (p. 14) overuse passive voice. Suggests how writers can learn to recognize passive voice in order to use it correctly.

765. Stohrer, Freda F. "Style in Technical Writing." Teaching English in the Two-Year College 7 (1981): 217-22.
Describes problems that occur from excessive use of passive voice.

766. Tebeaux, Elizabeth. "What Makes Bad Technical Writing

Bad? A Historical Analysis." IEEE--Transactions on Professional Communications PC-23 (1980):   71-76.
Explains that the use of "natural" word order in sentences will help reduce instances of bad technical writing.

767.   Van Ness, H. C.   "Technical Prose: English or Tech-lish?"   Chemical Engineering Education 11 (1977):   154-59, 173.
Examines the use of wordy, awkward constructions in technical prose and offers suggestions for eliminating such constructs from writing.

768.   Whittaker, Della A.   "Dephrase: Use Verbs."   Journal of Technical Writing and Communication 9 (1979):   59-67.
Provides sample sentences to show how the use of too many phrases and not enough verbs makes the meanings of the sentences unclear.   Suggests that technical writers "limit phrases and strengthen verbs" (p. 66) so that readers more easily understand what sentences mean.

8.6 Paragraphs

769.   Colby, John B.   "Paragraphing in Technical Writing." IEEE--Transactions on Professional Communications PC-20 (1977):   20-23.
Describes the characteristics of effective paragraphs.

770.   "How to Polish Your Writing."   IEEE--Transactions on Professional Communications PC-24 (1981):   179-81.
Discusses the importance of producing good paragraphs to communicate effectively.

771.   Jordan, Michael P.   "Some Associated Nominals in Technical Writing."   Journal of Technical Writing and Communication 11 (1981):   251-64.
Examines the use of "associated nominals," or noun phrases, to produce coherence in technical writing. Discusses some to "the types of meanings that can exist between clauses and sentences" (p. 261).   Suggests how existing studies of coherence can be used by the technical writing teacher to help students learn how to make their writing more coherent.

772. Marshek, Kurt M. "Transitional Devices for the Writer."
      IEEE--Transactions on Professional Communications PC-
      18 (1975): 320-22.
           Discusses the effective use of transitions to create
      coherence in writing.

773. McCanless, George F., Jr. "The Engineer and the
      Paragraph." IEEE--Transactions on Professional Com-
      munications PC-22 (1979): 198-99.
           Lists twenty-four checks that a writer can use to
      evaluate the effectiveness of paragraphs.

774. Selzer, Jack. "Another Look at Paragraphs in Techni-
      cal Writing." Journal of Technical Writing and Com-
      munication 10 (1980): 293-301.
           Argues that technical writers should pay attention
      to style at the paragraph level. Examines assumptions
      that lead to the disregard for paragraphs and explains
      why discussions of style should "attend to matters
      beyond the sentence" (p. 296).

8.7 Organization

775. Arnold, William R. "Vehicles for Documentation." IEEE
      --Transactions on Professional Communications PC-27
      (1984): 222-29.
           Explains how "vehicles," such as headings, bullets,
      figures, etc., can be used to improve the readability
      level of documents.

776. Brent, Douglas. "Indirect Structure and Reader Re-
      sponse." The Journal of Business Communication 22.2
      (1985): 5-8.
           Argues that using "buffer" sentences to make read-
      ers more receptive to "bad-news messages" and sales
      pitches more often irritates readers than appeases them
      by trying to "manipulate" readers' responses instead
      of matching readers' natural responses. Suggests that
      writers consider the entire rhetorical situation when
      deciding whether to use "indirect" or "direct" structure.

777. Girill, T. R. "Technical Communication and History."
      Technical Communication 32.4 (1985): 68-69.
           Examines the use of narrative in technical writing.

Finds that using narratives can cause numerous problems.

778. Gross, Alan G. "Style and Arrangement in Scientific
      Prose: The Rules Behind the Rules." Journal of Tech-
      nical Writing and Communication 14 (1984):   241-53.
      Presents the findings of a study made of files com-
      piled during the production of one issue of a scientific
      journal. Discusses changes found in style and arrange-
      ment used to present information within the articles.

779. Jordan, Michael P. "If Not Or But--Conjunctions in
      Sentential and Deductive Logic." Journal of Technical
      Writing and Communication 5 (1975):   23-38.
      Provides an introductory discussion of technical
      reasoning. Shows how the use of conjunctions such as
      "and" and "but" reflect the logic used in a technical
      document.

780. Lipson, Carol S. "Theoretical and Empirical Considera-
      tions for Designing Openings of Technical and Business
      Reports." The Journal of Business Communication 20.1
      (1983):   41-53.
      Discusses the use of redundancy in openings of
      business and technical reports. Shows that too much
      repetition of information is ineffective. Suggests ways
      to design effective openings.

781. Pinelli, Thomas E.; Virginia M. Cordle; and Raymond
      F. Vondran. "The Function of Report Components in
      the Screening and Reading of Technical Reports."
      Journal of Technical Writing and Communication 14
      (1984):   87-94.
      Presents the results of a survey conducted, in part,
      to determine how engineers and scientists actually read
      report components.

782. Souther, James W. "What to Report." IEEE--Transac-
      tions on Professional Communications PC-28 (1985):   5-8.
      Describes a study conducted to find out what in-
      formation managers want in reports and how managers
      extract that information from reports.

783. Swift, Marvin. "Writing a Conclusion." ABCA Bulletin
      44.4 (1981):   19-21.

Explains why conclusions to "problem-centered reports in business and industry" (p. 19) are hard to write. Offers guidelines for choosing terminology and organizational patterns that produce effective conclusions.

784. "The Process Model of Document Design." IEEE--Transactions on Professional Communications PC-24 (1981): 176-78.
Provides a model for the document design process.

785. Zappen, James P. "Writing the Introduction to a Research Paper: An Assessment of Alternatives." Technical Writing Teacher 12 (1985): 93-101.
Explains that three approaches exist for designing an introduction to a scientific or technical report. Evaluates the advantages and disadvantages of each approach.

786. Zimmerman, Muriel. "The Use of Repetition in Technical Communication." IEEE--Transactions on Professional Communications PC-25 (1983): 9-10.
Shows effective use of repetition in written documents.

• 9. WRITING ASSIGNMENTS •

9.1 Abstracts and Summaries

787. McGirr, Clinton. "Guidelines for Abstracting." Technical Communication 25.2 (1978): 2-5.
Explains how to write good abstracts.

788. Ratteray, Oswald M. T. "Expanding Roles for Summarized Information." Written Communication 2 (1985): 457-72.
Discusses seven types of summaries.

789. Roundy, Nancy. "A Process Approach to Teaching the Abstract." ABCA Bulletin 45.3 (1982): 34-38.
Explains how to teach students to write both descriptive and informative abstracts.

790. Waters, Max L. "Abstracting--An Overlooked Management Writing Skill." ABCA Bulletin 45.2 (1982): 19-21.
States the importance of teaching abstract writing in the business and technical writing classroom.

791. Weightman, Frank C. "The Executive Summary: An Indispensable Management Tool." ABCA Bulletin 45.4 (1982): 3-5.
Provides guidelines for writing effective executive summaries.

9.2 Case Problems

792. Catron, Douglas M. "A Case for Cases." ABCA Bulletin 47.1 (1984): 21-25.
Describes what constitutes effective and ineffective cases. Presents a sample case and writing assignment based on the case.

793. Kotler, Janet. "I Rest My Case." The Journal of
      Business Communication 22.4 (1985): 59-65.
         Describes a case assignment used in a business com-
      munication classroom to help students learn to under-
      stand others' points of view.

794. Stull, James B. "Solving Business Communication Prob-
      lems: A Journalistic Approach." ABCA Bulletin 46.3
      (1983): 10-11.
         Discusses a sample business communication problem
      to show how the journalistic method of determining
      "who?, what?, why?, where?, when?, and how?" can
      be used to solve the problem.

795. Weeks, Francis W. "How to Write Problems." ABCA
      Bulletin 41.2 (1978): 20-22.
         Discusses the development of case problems to be
      used for assignments in business communication courses.

796. York, Bonnie Brothers. "Candipop Incorporated: A
      'Casebook' Exercise." ABCA Bulletin 46.2 (1983): 28-
      31.
         Presents a case study to be used to generate writ-
      ing assignments. Provides sample assignments and
      "notes" for instructors.

9.3 Group Assignments

797. Cogin, William. "A Hands-On Project for Teaching In-
      structions." Technical Writing Teacher 8 (1980): 7-9.
         Describes an in-class assignment which consists of
      having students work in groups to write instructions
      for assembling and operating a race car set. Finds
      that the assignment helps students better understand
      the concepts of purpose, audience, and format.

798. Dobler, G. Ronald. "Teaching the Process Theme to
      Freshman Technical Writing Students Through Field
      Experiences." Technical Writing Teacher 5 (1977):
      22-25.
         Describes a process theme assignment that involves
      students working in pairs, interviewing merchants in
      the community, and writing process papers and thank-
      you letters.

799. Kogen, Myra. "In-Basket: Communication Through
      Classroom Dynamics." ABCA Bulletin 46.1 (1983): 15-
      18.
          Describes a classroom exercise designed to simulate
      "real-world" feedback to communication attempts by
      having students role play and respond to each other
      in writing.

800. Link, Thomas. "A Stressful Senior Project: A Stu-
      dent's View of Realistic Writing Assignments." Tech-
      nical Writing Teacher 10 (1983): 91-93.
          Describes the author's experiences with a writing
      project that failed because the co-writer for the project
      refused to write. Explains how working on such pro-
      jects, whether successful or not, benefits technical
      writing students.

801. Potvin, Janet H. "Using Team Reporting Projects to
      Teach Concepts of Audience and Written, Oral, and
      Interpersonal Communication Skills." IEEE--Transactions
      on Professional Communications PC-27 (1984): 130-37.
          Describes the use of team reporting projects at The
      University of Texas at Arlington.

802. Reep, Diana. "Questionnaires and Informal Reports:
      A Project for Freshman Technical Writing Students."
      Technical Writing Teacher 7 (1980): 74-75.
          Describes a project that consists of placing students
      in groups, having each group develop a questionnaire
      on a topic of choice, use the questionnaire to interview
      ten people, and write a four- or five-page informal re-
      port on the project.

803. Skarzenski, Donald. "A Problem-Solving Case for Tech-
      nical Writing Courses." Technical Writing Teacher 5
      (1978): 97-98.
          Describes a group assignment which presents groups
      with a business problem and asks the groups to write
      memos that offer possible solutions to the problems.

804. Wohlgamuth, William L., and others. "Do Structured
      Interpersonal Communication Activities Have an Effect
      on Students' Business-Writing Skills?" The Journal of
      Business Education 57 (1981): 66-69.
          Shows how group assignments can be used to im-
      prove students' writing skills.

9.4 Manuals and Specifications

805. Atlas, Marshall A. "The User Edit: Making Manuals
      Easier to Use." IEEE--Transactions on Professional
      Communications PC-24 (1981): 28-29.
         Suggests that the best way to test the effectiveness
      of a manual is to have a user follow the manual while
      performing the process the manual details.

806. Balog, Steven E., and Vern Wolfe. "The Mechanical
      Pencil as a Subject for Teaching Report Writing."
      Technical Communication 31.4 (1984): 58.
         Describes how a mechanical pencil can be used as
      the subject of a user's manual.

807. Berry, Elizabeth. "How to Get Users to Follow Pro-
      cedures." IEEE--Transactions on Professional Com-
      munications PC-25 (1982): 22-25.
         Describes the characteristics of well-written pro-
      cedures.

808. Bethke, Frederick J. "Technical Writing: Weaving the
      Silk Purse." Journal of Technical Writing and Com-
      munication 6 (1976): 263-67.
         Describes problems that are found in poorly designed
      technical manuals. Suggests techniques that will help
      writers improve manuals.

809. Dobrin, David. "Do Not Grind Armadillo Armor in This
      Mill." IEEE--Transactions on Professional Communica-
      tions PC-28 (1985): 30-37.
         Presents sample instructions to show what constitutes
      poorly written instructions. Suggests methods for
      writing good instructions.

810. Fleischhauer, F. William. "Technical Writing in the
      Management-Systems World." IEEE--Transactions on
      Professional Communications PC-28 (1985): 31-33.
         Describes a structure developed by the Grumman
      Corporation to improve the writing of procedures.

811. Funkhouser, R. Gay. "Functional Evaluation of Con-
      sumer Documents." The Journal of Business Communi-
      cation 20.3 (1983): 60-71.
         Discusses the importance of evaluating consumer

documents according to their functions. Describes a
study conducted to evaluate the "usefulness of contracts."

812. Gleason, James P., and Joan P. Wackerman. "Manual
      Dexterity--What Makes Instructional Manuals Usable."
      IEEE--Transactions on Professional Communications PC-
      27 (1984):   59-61.
          Discusses the components of effective instructional
      manuals.

813. Goldfarb, Stephen M. "Writing Policies and Procedures
      Manuals." IEEE--Transactions on Professional Com-
      munications PC-25 (1982):   14-15.
          Presents guidelines for writing good policies and
      procedures manuals.

814. Gudknecht, Alan R. "Well-Written Documentation Leaves
      Nothing to the Imagination." IEEE--Transactions on
      Professional Communications PC-25 (1982):   112-19.
          Describes the characteristics of good product docu-
      mentation.

815. Hackos, Joann T. "Using Systems Analysis Techniques
      in the Development of Standards and Procedures."
      IEEE--Transactions on Professional Communications PC-
      28 (1985):   25-30.
          Examines the "state" of standards and procedure
      writing. Suggests ways to produce effective standards
      and procedures.

816. Joseph, Albert. "Writing Training Materials That Turn
      People On." IEEE--Transactions on Professional Com-
      munications PC-24 (1981):   169-71.
          Offers suggestions for how to write useful training
      manuals.

817. Lay, Mary M. "Procedures, Instructions, and Specifi-
      cations: A Challenge in Audience Analysis." Journal
      of Technical Writing and Communication 12 (1982):
      235-42.
          Discusses the differences between procedures, in-
      structions and specifications in terms of the audience
      who use the documents. Thus, having students study
      and write these types of documents better trains stu-

dents for analyzing audiences in any communication situation.

818. Lopinto, Lidia. "Designing and Writing Operating Manuals." IEEE--Transactions on Professional Communications PC-27 (1984): 29-31.
   Suggests guidelines for designing and writing operating manuals.

819. Losano, Wayne A. "A Management-Run Orientation Program for Manual Writers." Journal of Technical Writing and Communication 6 (1976): 231-37.
   Suggests techniques that will help manual writers produce better documents.

820. Pilditch, James. "Did You Read the Instructions Carefully?" IEEE--Transactions on Professional Communications PC-24 (1981): 185-86.
   Stresses the characteristics of effective instructions for use of a product.

821. Ridgway, Lee. "The Writer as Market Researcher." Technical Communication 32.1 (1985): 19-22.
   Explains how writers of instruction and reference manuals for product must perform a "market analysis" to determine the needs of audiences for the manuals. Suggests some strategies for collecting such data.

822. Vink, Jamie. "Procedure Writing and Corporate Management: Thinking, Writing, and Speaking for Internal Coordination." Journal of Technical Writing and Communication 13 (1983): 394-53.
   Discusses the relationship between procedure manuals and the internal functions of an organization.

9.5 Memos, Letters, Brochures

823. Addison, Jim. "Brochures: A Teaching Rhetoric." Technical Writing Teacher 10 (1982): 21-24.
   Discusses the use of rhetorical principles in brochures. Provides a method for teaching brochures.

824. Ashdown, Paul G. "A Newsletter Writing Project in an Industrial and Technical Writing Course." Technical

Writing Teacher 2 (1975):   12-13.
Explains what technical writing students can learn
by producing newsletters--from market analysis stage
to distribution stage--for their major fields of study.

825.   Bell, James D.   "'Sell' Business Communication: Develop Community Contracts!"   ABCA Bulletin 46.3
(1983):   38-41.
Presents ways that students can apply instructions
for producing sales letters to real-world situations.

826.   Dolecheck, Carolyn Crawford.   "My Favorite Assignment: Communication Students--Sales Letter Consul-
tants for Small Business Firms."   ABCA Bulletin 43.1
(1980):   9-10.
Describes a letter assignment which consists of having students work with businesses to produce "real-
life" documents.

827.   Golen, Steven.   "Writing Actual Goodwill Product or
Service Letters: Putting Realism into Your Letter Assignments."   ABCA Bulletin 46.3 (1983):   31-32.
Describes an assignment which requires students to
write goodwill, product or service letters and actually
send these letters to companies.

828.   Harty, Kevin J.   "Some Guidelines for Saying 'No.'"
ABCA Bulletin 44.4 (1980):   23-25.
Suggests guidelines for writing letters that present
negative messages.

829.   Jacob, Nora B.   "Writing for Success in Business."
IEEE--Transactions on Professional Communications PC-
24 (1981):   121-22.
Explains how to write effective letters and memos.

830.   Potvin, Janet H.   "Eight Steps to Better Newsletters."
IEEE--Transactions on Professional Communications PC-
25 (1982):   204-10.
Presents an approach used to produce better newsletters.

831.   Prigge, Lila L.   "Students Develop Skills by Analyzing
Business Latters."   Business Education Forum 35.2
(1980):   26-27.

Explains how having students analyze and rewrite actual business letters collected from local businesses helps students better understand the importance of effective letter writing. Gives a sample check list that can be used to evaluate letters. Also, provides sample letters that have actually been used in the classroom.

832. Reep, Diana C. "The Manager's Message--Teaching the Internal Memo." ABCA Bulletin 47.1 (1984): 11-13.
Describes the major components of the internal memo. Suggests guidelines for writing internal memos. Details strategies for teaching students how to write effective internal memos.

833. Stoddard, Ted D. "The 'Real' Letter." ABCA Bulletin 48.2 (1985): 28-29.
Describes an assignment that requires each student to write a business letter and then mail the letter.

834. Thierfelder, William R., III. "The Misused Memo: Diagnosis and Treatment." Journal of Technical Writing and Communication 14 (1984): 155-62.
Explains how memos can be "misused." Discusses ways to produce useful memos.

835. Thompson, Isabelle. "The Job-Search Memo." ABCA Bulletin 48.3 (1985): 29-30.
Describes a memo assignment that aims to help students evaluate their possible employment opportunities.

836. Whittaker, Della A. "From Term Paper to Brochure." Teaching English in the Two-Year College 5 (1978): 45-46.
Describes an assignment which requires students to take information from their term papers and present this information in a brochure suitable for publication.

837. Wilcox, Alan D. "How to Write a Recommendation." IEEE--Transactions on Professional Communications PC-27 (1984): 211-14.
Suggests guidelines for writing recommendation letters.

9.6 Proposals

838. Beck, Clark E. "Proposals: Write to Win." IEEE--

Transactions on Professional Communications PC-26
(1983):   56-57.
  Suggests ways to write a winning proposal.

839. Corey, Robert L.   "The Persuasive Technical Proposal:
      Rhetorical Form and the Writer." Technical Communica-
      tion 22.4 (1975):   2-5.
        Explains how to make proposals persuasive.

840. Flechtner, Adalene Smith.   "BIOS Can Make Your Pro-
      posal a Winner." Technical Communication 24.1 (1977):
      4-6.
        Explains how to write biographical sketches to be
      used in proposals.

841. Gross, Gerald J.   "Work Together, Write Together:
      Group Projects in Technical Writing."   Conference on
      CCC, Kansas City, Missouri, 31 March-2 April 1977.
        Describes a group proposal assignment.

842. Horobetz, Joseph S.   "'No Chapter on Proposal Writing'
      Does Not Equal 'No Proposal Writing.'" Technical
      Writing Teacher 5 (1978):   95-96.
        Describes a proposal assignment which requires
      students to write and submit a proposal that could
      actually be used outside of the classroom.

843. Kennedy, George E.   "Teaching Formal Proposals: A
      Versatile Minicourse in Technical Writing." Journal of
      Technical Writing and Communication 13 (1983):   123-
      37.
        Details all considerations required in proposal writing.

844. Minor, Dennis E.   "Teaching the Technical Writing
      Proposal." Technical Writing Teacher 7 (1979):   24-27.
        Discusses a proposal assignment that consists of
      each student proposing the long report that he/she
      will do during the course. Within the proposal, the
      student must show the feasibility of the project as well
      as a procedure for completing the project.

845. Myers, Greg.   "The Social Construction of Two Bio-
      logists' Proposals." Written Communication 2 (1985):
      219-45.
        Compares versions of research proposals written by

two biologists to show what changes they made to suit differing audiences.

846. Norman, Rose, and Marynell Young. "Using Peer Reviews to Teach Proposal Writing." Technical Writing Teacher 12 (1985): 1-9.
Suggests a series of assignments to teach students how to write a formal proposal by "imitating the grant proposal and review process" (p. 2). Believes that by using preliminary writing assignments along with oral reports and peer reviews helps students develop good topics and write good formal proposals.

847. Pittendrigh, Adele. "The $500.00 Proposal." Exercise Exchange 28.2 (1983): 16-19.
Describes a grant-proposal writing assignment.

848. Seisler, Jeffrey M. "Proposal Writing: Approaching the Approach." IEEE--Transactions on Professional Communications PC-26 (1983): 58-59.
Discusses the contents of a good approach section in a proposal.

849. Smith, Frank R. "Education for Proposal Writers." Journal of Technical Writing and Communication 6 (1976): 113-22.
Describes what is required of proposal writers in the aerospace industry. Suggests ways to better prepare technical writing students for proposal writing tasks.

850. _____. "Use Rhetoric for Better Technical Proposals." Technical Communication 22.3 (1975): 12-13.
Suggests a quick method for polishing the rhetoric of technical proposals.

851. Whalen, Tim. "Grant Proposals: A Rhetorical Approach." ABCA Bulletin 45.1 (1982): 36-38.
Explains how to produce successful proposals.

852. _____. "Renewal: Writing the Incumbent Proposal." IEEE--Transactions on Professional Communications PC-28 (1985): 13-16.
Explains the importance of collecting pertinent data when writing RFPs, lobby reports, documented personnel assessments, and vulnerability assessment reports.

9.7 Publication Process

853.   Estrin, Herman A.   "Writing for Publication."   Journal
       of Technical Writing and Communication 5 (1975):   99-
       102.
           Discusses the benefits of writing for publication.
       Provides guidelines for evaluating journals' styles and
       for following protocol when submitting manuscripts.

854.   Gould, Jay R.   "The Letter of Query."   Journal of
       Technical Writing and Communication 5 (1975):   107-17.
           Explains why one should submit a letter of inquiry
       before submitting manuscripts for publication.   Offers
       guidelines for writing letters of inquiry.

855.   Holder, Fred W.   "Try Writing a Technical Article."
       IEEE--Transactions on Professional Communications PC-
       19 (1976):   38-41.
           Explains the steps involved in researching, writing,
       and publishing an article.

9.8 Reports

856.   Annett, Clarence H.   "A Problem-Solving Model for
       Organizing Short Technical Reports."   Technical Com-
       munication 29 (1982):   11-14.
           Describes the nature of short reports.   Provides an
       eight-step model for solving problems and producing
       written reports based on the model.

857.   Atkinson, Ted.   "My Favorite Assignment: Doing In-
       stitutional Research to Teach the Long Report."   ABCA
       Bulletin 46.1 (1983):   10-11.
           Describes report assignments that involve students
       working in conjunction with offices of institutional re-
       search.

858.   Brown, Judith C.   "Controlled Walkthrough Approach
       for Analytical Report Writing."   ABCA Bulletin 46.3
       (1983):   35-37.
           Shows how the concept of "controlled walkthrough"
       used by "leaders" in data processing can be adapted
       to a strategy for report writing.

859. Day, Robert A. "How to Write a Scientific Paper."
      IEEE--Transactions on Professional Communications
      PC-20 (1977): 32-37.
         Presents the components of and the correct presenta-
      tion of information in a scientific paper.

860. Driskill, L. P. "Motivating Students with a Winning
      Assignment." Journal of Technical Writing and Com-
      munication 7 (1977): 129-34.
         Describes how a major writing assignment, a feasi-
      bility study, is effectively used at Rice University.
      Explains the steps that students follow in order to
      choose topics, analyze the audiences, write the reports,
      and give oral presentations of the report. States some
      possible topics for the reports.

861. Fisher, John K., and Ted Atkinson. "Writing Surveys
      for Institutional Research: A Profitable Collaboration."
      ABCA Bulletin 46.2 (1983): 34-37.
         Provides guidelines that will help students write
      surveys to collect data for an institutional research
      project.

862. Frayer, Bill. "The Feasibility Report: Teaching Re-
      port Writing at a Two-Year Technical Institute." Tech-
      nical Writing Teacher 10 (1983): 195-99.
         Describes a fifteen-week class project that involves
      not only researching and writing a feasibility report
      but also orally presenting a progress report and a
      shortened version of the feasibility report. Explains
      why the feasibility report project is "an ideal assign-
      ment for two-year technical students at the associate
      degree level" (p. 195).

863. Friedman, Sharon M. "Using Real World Experience to
      Teach Science and Environmental Writing." Annual
      Meeting of the Association for Education in Journalism,
      Seattle, 13-16 August 1978.
         Stresses the importance of students' receiving "real-
      world" writing experience. Describes a report-writing
      assignment used at Lehigh University.

864. Fuller, Don. "How to Write Reports That Won't Be
      Ignored." IEEE--Transactions on Professional Com-
      munications PC-23 (1980): 79-81.

Emphasizes audience awareness and clarity as key components of good report writing.

865. Harty, Kevin J.   "Campus Issues: A Source for Research in Business and Technical Writing Courses." Teaching English in the Two-Year College 12 (1985): 221-24.
Examines the importance of report writing and suggests how report projects can be devised from "campus issues."

866. Hissong, Douglas W.   "Write and Present Persuasive Reports."   IEEE--Transactions on Professional Communications PC-21 (1978):   150-52.
Provides guidelines to help writers produce good persuasive reports.

867. Johnston, Sue Ann.   "From Shakespeare to the Nuts and Bolts of Writing."   English Quarterly 16.4 (1984): 26-28.
Describes a formal report assignment designed to increase students' audience awareness.

868. Jordan, Michael P.   "Structured Information in Functional Writing."   Teaching English in the Two-Year College 9 (1982):   61-64.
Examines various structures which can be used to organize information for report writing.

869. Kaftan, Robert A.   "The Long Report: A Comprehensive Model."   ABCA Bulletin 44.4 (1981):   35-36.
Explains how the long report assignment can be expanded to include a proposal assignment, a progress report assignment, and oral presentation assignments.

870. Killingsworth, M. Jimmie.   "The Essay and the Report: Expository Poles in Technical Writing."   Journal of Technical Writing and Communication 15 (1985):   227-33.
Examines the roles of essays and reports in expository writing.   Discusses the relationship between these two types of writing.

871. Lewis, William J.   "Two Technical Writing Assignments: Teaching Ideas."   English Journal 67.4 (1978):   65-68.
Describes assignments that require students to write a product research report and a survey report.

872. Macchi, Reynolds. "The Inadequacies of Technical Re-
     porting." Journal of Technical Writing and Communica-
     tion 6 (1976): 255-58.
          Discusses improvements that need to be made in the
     area of technical reporting.

873. McLaren, Margaret C. "Useful Reporting: Linkspan
     Between University and Employment." ABCA Bulletin
     47.3 (1984): 47-49.
          Discusses advantages of requiring students to write
     investigative formal reports in conjunction with organiza-
     tions outside the university.

874. Minor, Dennis E. "A Structure for the Problem-Solving
     Paper." Technical Writing Teacher 10 (1982): 8-11.
          Provides sample structures for setting up a problem-
     solving paper.

875. "My Favorite Assignment." ABCA Bulletin 40.3 (1977):
     19-20.
          Describes an assignment that requires students to
     plan a conference and write accompanying reports.

876. Rothmel, Steven Zachary. "The Student As Consultant."
     Journal of Technical Writing and Communication 9 (1979):
     311-15.
          Describes a formal report assignment that requires
     students to place themselves in the roles of consultants.

877. Roundy, Nancy. "Structuring Effective Technical Re-
     ports." Technical Communication 32.1 (1985): 26-29.
          Provides step-by-step guidelines for entirely struc-
     turing a report.

878. Ruch, William V. "The Corporate Annual Report as a
     Teaching Aid in Business Communication Classes."
     ABCA Bulletin 42.4 (1979): 1-3.
          Explains how annual reports can be used for assign-
     ments in the technical writing classroom.

879. Smith, Denise M., and Nick L. Smith. "Writing Ef-
     fective Evaluation Reports." Studies in Educational
     Evaluation 7 (1981): 33-41.
          Offers guidelines for producing evaluation reports.

880. Southard, Sherry. "Humanistic Research Projects:
      The Basis for a Technical Report." Annual Meeting of
      the Midwest Section of the American Society for Engi-
      neering Education. Wichita, Kansas, 23 March 1984.
          Describes a report assignment used to teach students
      how to use the scientific method for researching and
      writing reports.

881. Swift, Marvin. "Writing a Problem-Centered Report."
      ABCA Bulletin 46.3 (1983):   19-23.
          Describes methods for collecting data and writing a
      report in industry.

882. Thomas, Edward G. "The Corporate Annual Report:
      A Basic Resource in the Written Communications Course."
      ABCA Bulletin 41.2 (1978):   29-34.
          Examines the use of annual reports to generate as-
      signments in business writing courses.

883. Urlich, Gael D. "Write a Good Technical Report."
      IEEE--Transactions on Professional Communications
      PC-27 (1984):   14-19.
          Provides guidelines for writing technical reports.

884. Van Oosting, James. "The 'Well-Made' Report." ABCA
      Bulletin 45.4 (1982):   9-10.
          Discusses the components that make up an effective
      report.

885. Waltman, John L. "Towards More Manageable Formal
      Report Assignments." ABCA Bulletin 46.4 (1983):   29-
      30.
          Describes a formal report assignment used at Louisi-
      ana State University.

886. White, Roberta D. "My Favorite Assignment: A Mean-
      ingful Unit in Business Communication." ABCA Bulletin
      43.2 (1980):   22-23.
          Describes an assignment that requires students to
      research a corporation and produce a formal report
      that discusses the results of the research.

887. Young, Art. "What's Really Basic About Usage in Tech-
      nical Writing?" Technical Writing Teacher 4 (1977):
      68-72.

Suggests adapting the formal report, written mainly in businesses, to the Freshman English classroom in order to teach students to consider "the voice of the writer, the audience addressed, the subject matter to be written on, and the purpose for which it is written" (p. 69).

888. Zins, Joseph E., and David W. Barnett. "Report Writing: Legislative, Ethical, and Professional Challenges." Journal of School Psychology 21 (1983): 219-27.
    Examines changes in all areas of psychological report writing. Provides a new approach to report writing that attempts to address these changes.

9.9 The Job Search

889. Bernheim, Mark. "The Written Job Search--Doubts and 'Leads.'" Technical Writing Teacher 10 (1982): 3-7.
    Examines areas of concern associated with the job search. Suggests ways to obtain "leads" on possible employment opportunities.

890. Blicq, Ron S. "Job Hunting: Sharpening Your Competitive Edge." IEEE--Transactions on Professional Communications PC-24 (1984): 201-10.
    Discusses resume writing, letter of application, employment application forms, and interviews.

891. Dittman, Nancy A. "Job Winning Résumés." ABCA Bulletin 46.2 (1983): 16-22.
    Examines various types of résumés and encourages teachers to be more flexible when teaching students how to design résumés for various job opportunities. Includes sample résumés.

892. Egan, Christine. "Writing Résumés and Cover Letters." IEEE--Transactions on Professional Communications PC-24 (1981): 156-60.
    Discusses the importance of self-analysis and market analysis to the job hunt. Explains how to write good résumés and cover letters.

893. Fox, Marcia R. "The Cover Letter." IEEE--Transactions

on Professional Communications PC-24 (1981):   163-65.
Explains how to write a good cover letter.

894. Hunt, Bridgford. "The New You--Researched, Résuméd,
and Rarin' to Go." IEEE--Transactions on Professional
Communications PC-24 (1981):   160-62.
Discusses writing résumés and interviewing.

895. Hutchinson, Kevin L. "Personnel Administrators' Pref-
erences for Resume Content: A Survey and Review of
Empirically Based Conclusions." The Journal of Business Communication 21.4 (1984):   5-14.
Describes a survey of five hundred personnel administrators at Fortune 500's top-ranked companies to
learn what they considered to be important for résumés.

896. Penrose, John M. "A Discrepancy Analysis of the Job-
Getting Process and a Study of Résumé Techniques."
The Journal of Business Communication 21.3 (1984):
5-15.
Describes a study conducted to compare recruiters'
views and students' views concerning interviews and
résumés.

897. Trace, Jacqueline. "Teaching Resume Writing the Functional Way." ABCA Bulletin 48.2 (1985):   34-41.
Examines the advantages of using a functional résumé.
Describes a workshop that can be used when teaching
students how to design a functional résumé.

898. Watkins, Donna M. "Creating 'What I Can Do for You'
Emphasis in Application Letters." ABCA Bulletin 44.4
(1980):   6-7.
Suggests ways that students can improve their letters
of application.

899. _____. "Sales Letters and Application Letters: Draw-
ing the Parallel." ABCA Bulletin 46.2 (1983):   32-33.
Discusses similarities that exist between writing sales
letters and writing application letters.

9.10 Various Assignments

900. "An Assignment for All Seasons." Technical Writing

Teacher 7 (1980):   60-63.
Describes an assignment which consists of having each student "describe a scholarly journal in the student's major field according to ... questions" devised by the instructor (p. 60). Explains how this assignment meets the needs of a heterogeneous technical writing classroom.

901.   Arnold, Vanessa Dean. "Teaching Nonverbal Behavior." ABCA Bulletin 46.2 (1983):   27.
Describes an assignment that asks each student to write a paper describing nonverbal messages sent by two instructors whom the student has observed.

902.   Atkinson, Ted. "Politics and the Business Writing Student: An Approach to Finding Real Writing Projects." ABCA Bulletin 45.4 (1982):   11-12.
Presents writing assignments which consists of students working in conjunction with their state senator.

903.   Bielawski, Larry. "Process Writing as an Ice-Breaker." Technical Writing Teacher 11 (1984):   120-21.
Describes a process assignment to be given on the first or second day of class. Finds that beginning the semester with this assignment helps students to quickly grasp the concepts of context and audience.

904.   Britton, W. Earl. "Organizing the Technical Description." Technical Communication 29 (1982):   14-15.
Finds that the main problems of poorly written technical descriptions stems from poor organization of information.

905.   Ceccio, Joseph F., and Michael J. Rossi. "Inventory of Students' Sex Role Biases: Implications for Audience Analysis." Technical Writing Teacher 8 (1981): 78-82.
Describes an exercise designed to help students recognize their own sex role biases and "to help them to recognize their need to generalize accurately" (p. 78).

906.   Clampitt, Philip G. "Effective Performance Appraisal: Viewpoints from Managers." The Journal of Business Communication 22.3 (1985):   49-57.

Discusses the findings of a survey of managers to
determine how they viewed existing "schemes" of performance appraisals. Examines four problems that
managers found with existing schemes.

907. Clinkscale, Bella G. "Model for Multi-Faceted Assignment Planning in Oral Communication." ABCA Bulletin
47.2 (1984): 27-31.
Describes a business communication assignment that
requires students to use various communication skills.

908. Corey, Jim. "Assignment: Journal Analysis--A Para-
digm for Technical Writing." Technical Writing Teacher
8 (1981): 80-82.
Describes a journal analysis assignment which has
each student read several issues of one journal in his/
her field and then write a report that presents "valid
general conclusions about the journal" (p. 80). Provides a detailed outline for the journal analysis report.

909. Cottone, James F. "Writing an Invention Disclosure."
IEEE--Transactions on Professional Communications
PC-22 (1979): 105-08.
Presents guidelines for correctly writing an inven-
tion disclosure to be used when applying for a patent.

910. Craven, Gerald A. "Reworking the Foul Copy: An
Exercise in Revision." Technical Writing Teacher 4
(1977): 105-06.
Provides an exercise that students can use when
evaluating the term papers before revising them.

911. Dunagan, David. "Suggestions for an Advanced Technical Writing Course." Technical Writing Teacher 7
(1980): 119-21.
Examines textbooks and possible assignments.

912. Emerson, Frances B. "Applied Pragmatics in the Development of Technical Writing Assignments." Technical Writing Teacher 8 (1981): 63-68.
Explains how "at least three major points of American pragmatism ought to be applied to the development
of technical writing assignments" (p. 63). States that
teachers should create a practical environment, realis-
tic contexts, and verifiable assignments. Provides
sample assignments that teachers can use in the classroom.

913. Estrin, Herman A. "Motivating and Preparing Students
to Submit Articles on Technical Writing: Staffroom
Interchange." College Composition and Communication
27 (1976): 300-400.
States that in order to become better technical
writers themselves, technical writing students must
first become acquainted with journals that contain articles on technical writing. By reading what the articles
in such journals say about technical writing, students
come to understand more about the nature of technical
writing. Next, lists topics for discussion in the technical writing classroom. Offers an assignment that involves students reading articles on technical writing
that were written by other students.

914. _____. "The Teaching of Technical Writing: Fasci-
nating, Fulfilling, and Rewarding." Technical Writing
Teacher 10 (1983): 167-73.
Describes six writing projects used in a semester-
long technical writing course. Finds that these projects
help students understand various "facets of technical
writing styles" and encourages students to publish.

915. Freier, Martin. "Effective Specification Writing." Technical Communication 22.2 (1975): 14-16.
Discusses how to write good specifications.

916. Gilsdorf, Jeanette W. "High-Tech Update for a Business Communication Class." ABCA Bulletin 46.3 (1983):
13-14.
Describes an assignment that requires students to
compile a brief annotated bibliography on the use of
high-technical communication in business.

917. Goldstein, Jone Rymer, and Elizabeth L. Malone. "Using
Journals to Strengthen Collaborative Writing." ABCA
Bulletin 48.3 (1985): 24-28.
Discusses the value of journal writing assignments.

918. Gould, John D., and Stephen J. Boies. "Writing, Dic-
tating, and Speaking Letters." IEEE--Transactions on
Professional Communications PC-22 (1979): 16-18.
Describes an experiment conducted to study the
speed and quality of dictation as performed by both
"novice" and "experienced" dictators.

919.   Grodsky, Susan J.   "Indexing Technical Communications:
       What, When, and How."   Technical Communication 32.2
       (1985):   26-30.
       Examines the importance of indexing and provides
       guidelines for how to index.

920.   Herbert, Mellanie, and Mary Margaret Hosler.   "Office
       Style Dictation Simulation--Both Sides Can Benefit."
       ABCA Bulletin 47.1 (1984):   26-28.
       Describes a dictation exercise used to allow students
       to experience on the job dictating.

921.   Hine, Edward A.   "Your Technical Paper Won't Write
       Itself."   Chemical Engineering 82 (1975):   77-78.
       Provides guidelines for writing technical papers.

922.   Hogan, Michael.   "A Community-Based Project for Tech-
       nical Writing."   Technical Writing Teacher 6 (1979):
       99-102.
       Suggests that one way to make what is taught in
       technical writing classes relevant to students is to have
       students' assignments revolve around some issue that is
       important to the community.   Supports this suggestion
       by describing a report project that focused on the
       problem of duplicate city street names.

923.   Hunter, Paul.   "Corporate Irresponsibility and the
       Business Communication Course."   ABCA Bulletin 40.2
       (1977):   26-28.
       Draws from business practices to produce situational
       writing assignments.

924.   Kelly, Kathleen.   "Analyze a Communication Problem in
       the Workplace: A Report Assignment for the Part-Time
       MBA Student."   ABCA Bulletin 48.3 (1985):   30-33.
       Describes an assignment which requires students to
       "analyze a communication problem in the organization
       for which they work" (p. 30).

925.   Kelton, Robert W.   "Teaching Students to Organize
       Reviews of Literature."   Technical Writing Teacher 8
       (1981):   72-74.
       Defines review of literature and presents advantages
       of assigning reviews of literature to technical writing
       students.   Provides sample exercises for a review of

literature and analysis of the results students produce
from the exercise.

926. Kogen, Myra. "The College as Laboratory in Teaching
Business Communication." Teaching English in the Two-
Year College 10 (1984): 247-50.
Suggests that teachers can draw from the college
environment to create more "real-life" writing assign-
ments.

927. Kramer, Melinda G. "No Time to Write: Business Operating Procedures in the Classroom." ABCA Bulletin
42.4 (1979): 27-28.
States that dictation should be taught as a part of
the technical writing classroom.

928. McNair, John R. "Mystery Mechanism: A Pre-Writing
Exercise." Teaching English in the Two-Year College
8 (1981): 55-56.
Describes a pre-writing assignment which requires
students to learn about an unfamiliar mechanism through
a series of "clues."

929. Murphy, J. William. "What Are You Going to Be When
You Grow Up?" ABCA Bulletin 46.4 (1983): 30-31.
Presents a "progress" report assignment that re-
quires students to examine their career goals.

930. Porter, James E. " 'Instructions' to Introduce Technical Writing." Technical Writing Teacher 11 (1984):
116-17.
Describes a first-day assignment used to show students "how the rhetorical principles learned in freshman composition apply to technical writing and to render
the concept of 'technical writing' less threatening" (p.
116).

931. Ricks, Don M. "To Teach Writing You Must Teach
Success." Canadian Training Methods 8.3 (1975): 16-
17.
Explains why writing should be considered a behavior. Offers guidelines for effective writing assignments.

932. Scott, James Calvert. "The Pictogram." ABCA Bulletin

47.3 (1984):   42-43.
Describes a pictogram assignment to rekindle students' motivation in the business and technical writing
course.

933.   Senf, Carol. "The Portfolio or Ultimate Writing Assignment." Technical Writing Teacher 11 (1983):   23-25.
Suggests that students should compile a portfolio of
all assignments produced in the technical writing classroom. Describes how a teacher can incorporate the
portfolio assignment into the technical writing classroom.
Explains that presenting a portfolio during an interview
makes the student seem more professional.

934.   Soares, Eric J., and Leslie A. Goldgehn. "The Portfolio Approach to Business Communication." ABCA
Bulletin 48.3 (1985):   17-21.
Examines the use of requiring students to prepare
portfolios in the business communication course.

935.   Souther, James W. "Developing Assignments for Scientific and Technical Writing." Journal of Technical Writing and Communication 7 (1977):   261-69.
Discusses what constitutes a "good" assignment for
teaching aspects of scientific and technical writing.
Describes three assignments that can be used.

936.   Stalnaker, Bonny J. "Documented Position Paper: A
Useful Approach to Teaching the Formal Research Report." ABCA Bulletin 41.2 (1978):   15-17.
Describes a short written assignment to help students learn correct documentation procedures.

937.   Tebeaux, Elizabeth. "Getting More Mileage Out of Audience Analysis--A Basic Approach." Journal of Technical Writing and Communication 12 (1982):   15-24.
Explains how students can learn much about audience
analysis by selecting a passage from an advanced text
in their fields and rewriting this passage to two other
specified audiences.

938.   Varner, Iris I., and Carson H. Varner. "The Press
Release to Illustrate Reader Adaptation in Business Report Writing." ABCA Bulletin 42.3 (1979):   3.
Explains how having students write press releases
teaches students about audience awareness.

939.   Vaughn, Jeanette W., and others. "Teach Business
        Communication Students to Dictate--Both Ways." ABCA
        Bulletin 45.4 (1982): 20-22.
            Examines the importance of teaching students how to
        dictate.

940.   Vidali, Joseph J. "Using Critiques of Business Re-
        search for Practical Experience." The Journal of Busi-
        ness Education 52.3 (1976): 124-25.
            Discusses the advantages of having students write
        critiques of business research. Provides an outline
        that can be used when writing a critique.

941.   Wager, Inez, and Willis Wager. "Something More Than
        Communication." Technical Writing Teacher 9 (1982):
        179-83.
            Describes what hidden meanings can exist in some
        technical writing assignments.

942.   Walraven, Eddie. "Two Classroom Exercises in Audi-
        ence Adaptation." Technical Writing Teacher 4 (1977):
        55-58.
            Describes two exercises for improving students' skills
        in audience adaptation. The first exercise involves
        presenting an oral report and answering questions
        from the audience. The second consists of reading re-
        ports "ranging from the fairly technical to the highly
        complex" and then relating information presented in
        these reports to "everyday circumstances" (p. 57).

943.   Walzer, Arthur E. "Ethos, Technical Writing, and the
        Liberal Arts." Technical Writing Teacher 8 (1981):
        50-53.
            Believes that ethos is of major importance in the
        technical writing classroom. States that students should
        "analyze the ethos of the profession they are preparing
        to enter" (p. 50). Describes an assignment which asks
        students to examine trade journals to study the ethos
        of their intended profession.

944.   Whitburn, Merrill D. "The First Day in Technical Com-
        munication: An Approach to Audience Adaptation."
        Technical Writing Teacher 3 (1976): 115-18.
            Describes how the importance of audience adaptation
        is made apparent during the first day in a technical

communication class by having students give "impromptu"
talks about themselves and by assigning a letter of introduction for the students to write as homework.

945. White, Jane F., and others. "My Favorite Assignment."
      ABCA Bulletin 44.4 (1980): 20-22.
          Describes an assignment which includes library re-
      search and a process description writing assignment.

946. Whittaker, Della A. "Practice What's Practical." Tech-
      nical Writing Teacher 5 (1978): 91-92.
          Discusses technical writing assignments that are
      relevant to adult students taking a technical writing
      course.

947. _____. "Teaching Students to Prepare Their Own
      Style Guides." Technical Writing Teacher 6 (1979):
      103-05.
          Describes how to develop an assignment for having
      students "prepare a style guide for use at work" (p.
      103). Provides samples from style guides produced by
      students.

948. _____. "The Book Review: A Class Assignment."
      Teaching English in the Two-Year College 6 (1980):
      261-64.
          Describes a book review assignment that requires
      students to read reviews in professional journals and
      then submit their own reviews to a professional journal.

949. Amsden, Dorothy Corner, and Scott P. Sanders. "Developing Taste and Judgment: Correctness and the Technical Editor." Technical Writing Teacher 12 (1985): 111-14.

Provides a detailed example to show how "editors and writers routinely consult reference books as they work, and use what they find to exercise editorial judgment" (p. 113). Suggests that such an example can be effectively used in a technical editing classroom.

950. Annett, Clarence H. "Improving Communication: Eleven Guidelines for the New Technical Editor." Journal of Technical Writing and Communication 15 (1985): 175-79.

Examines seven guidelines that a new technical editor can use when working with authors and editing texts.

951. Boomhower, E. F. "Producing Good Technical Communications Requires Two Types of Editing." Journal of Technical Writing and Communication 5 (1979): 277-81.

Explains the importance of performing both literary and technical editing in order to produce effective documents.

952. Bostian, Frieda F. "Elements of Style in Technical Writing Classes." Technical Writing Teacher 10 (1983): 124-25.

States that technical writing students need to know how to edit. Describes a method for teaching editing by using Strunk and White's Elements of Style as an editing guide.

953. Caernarven-Smith, Patricia. "Driving a Typesetter." Technical Communication 32.2 (1985): 50-53.

Examines the use of typesetters in technical publishing.

954.   Cederborg, Gibson A.   "The Role and Rationale of
       Technical Editors."   Journal of Technical Writing and
       Communication 5 (1975):   283-86.
       Explains what characteristics a good technical editor
       possesses.

955.   Chaffee, Patricia.   "Human Engineering and Technical
       Writing."   Technical Writing Teacher 10 (1983):   130-33.
       Discusses the relationship between the "technical
       editor and the engineer or scientist author" and de-
       scribes problems that can occur in this relationship.
       Provides methods for handling problems that arise from
       "such negative forces as ego, fear, insecurity, and
       defensiveness" (p. 130).   Suggests that students should
       edit each other's reports in order to understand the
       editor/author relationship.

956.   Cheney, Patrick, and David Schleicher.   "Redesigning
       Technical Reports: A Rhetorical Editing Method."
       Journal of Technical Writing and Communication 14
       (1984):   317-37.
       Explains how editors can use "rhetorical editing" to
       improve the overall effectiveness of reports.   Describes
       how to perform a rhetorical edit.

957.   Hageman, Mary S., and Louise M. Vest.   "A Monologue-
       Dialogue Workshop for Teaching Editing."   Technical
       Communication 32.4 (1985):   66.
       Describes a method for teaching effective communica-
       tion between writers and editors.

958.   Lewenstein, Bruce V.   "What You See Is What You Get:
       How, Not What, to Proofread."   Technical Communication
       32.1 (1985):   23-25.
       Provides guidelines for how to improve proofreading
       skills.

959.   Mann, Michele H.   "How to Edit the Passive Writer's
       Work."   Technical Communication 32.3 (1985):   14-15.
       Discusses the characteristics of a "passive writer."
       Suggests guidelines which will assist editors in helping
       passive writers become better writers.

960. Manusco, Joseph C. "Teaching Interview Strategies for Technical Editors." Technical Communication 32.1 (1985): 43.
Discusses strategies that technical editors can use when interviewing technical authors.

961. Masse, Roger E. "Theory and Practice of Editing Processes in Technical Communication." IEEE--Transactions on Professional Communications PC-28 (1985): 34-42.
Examines various approaches to the editing process.

962. Putnam, Constance E. "Myths About Editing." Technical Communication 32.2 (1985): 17-20.
Refutes nine "myths" concerning editing.

963. Soderston, Candace. "The Usability Edit: A New Level." Technical Communication 32.1 (1985): 16-18.
Discusses types of editing that can be performed on a written document. Suggests that a new type of editing, "usability editing," be added to the editorial process.

• 11. ORAL COMMUNICATIONS •

11.1 Teaching Oral Communication

964. Allen, Thomas. "Attentive Listening Is the Key to Effective Oral Business Communication." Balance Sheet 59 (1978): 292-93, 324.

Discusses the importance of listening in communication and describes common barriers to effective listening.

965. Davis, William L. "The Oral Approach to Business Communication." ABCA Bulletin 41.2 (1978): 9-14.

Discusses issues involved in developing a course in oral communication. Provides a sample course outline and sample assignments.

966. Geonetta, Sam C. "Increasing the Oral Communication Competencies of the Technological Student: The Professional Speaking Method." Journal of Technical Writing and Communication 11 (1981): 233-44.

Presents methods that can be used to improve students' oral communication skills.

967. Glossner, Alan J. "Improving Oral Reports: A Heuristic Approach." ABCA Bulletin 46.3 (1983): 32-34.

Discusses an oral presentation assignment that incorporates the use of videotaping.

968. Goldstein, Jone Rymer. "Dialogue: The Critical Oral Skill for Students in Technical Writing." Technical Writing Teacher 8 (1981): 54-58.

Stresses the need for teaching technical writing students not only how to give oral presentations but also how to conduct one-to-one dialogue. Explains why dialogue skills are important and how to incorporate the

teaching of dialogue skills in the technical writing classroom.

969. Hamermesh, Madeline. "Sharpening the Old Saws:
      Speech-Aact Theory and Business Communication."
      The Journal of Business Communication 18.2 (1981):
      15-22.
         Examines speech-act theory, describing the major
      points of the theory and showing what speech-act
      theory has in common with "other communication models."
      Explains what speech-act theory offers teachers of
      communication.

970. Hand, Harry E. "Technical Speech: A Need for Teach-
      ing and Research." IEEE--Transactions on Professional
      Communications PC-18 (1975): 18-21.
         Presents the results of studies conducted to deter-
      mine what kind of research and teaching needs to be
      done in the area of technical speech.

971. Hewing, Pernell Hayes. "A Practical Plan for Teaching
      Oral Communication in the Business Communication
      Course." ABCA Bulletin 40.4 (1977): 9-11.
         Discusses how to incorporate oral presentation assign-
      ments into a formal report writing assignment to improve
      students' speaking and writing skills.

972. Hulbert, Jack E. "Conducting Intelligent Business
      Dialogue." ABCA Bulletin 43.2 (1980): 3-6.
         Examines current teaching practices concerning oral
      communication. Suggests a way to create effective oral
      communication.

973. Johnson, Jack E. "My Favorite Assignment." ABCA
      Bulletin 45.4 (1982): 26-30.
         Describes the use of conferencing to increase stu-
      dent involvement in oral presentations.

974. Mayer, Kenneth R. "Developing Delivery Skills in
      Oral Business Communications." ABCA Bulletin 43.3
      (1980): 21-24.
         Presents tips to help teachers encourage students to
      develop good delivery skills for presenting oral reports.

975. Mitchell, G. A. "The Development of Oral Skills for
      the Presentation of Technical Information." Journal of

Technical Writing and Communication 14 (1984):   109-12.

Examines the use of public speaking exercises to develop oral presentation skills.

976.   Page, William T.   "Helping the Nervous Presenter: Research and Prescriptions."   The Journal of Business Communication 22.2 (1985):   9-19.

Evaluates the research that has been done to determine the causes of nervousness in presenters and points out the validity of some theories.   Describes remedies for nervousness, explaining that one must consider whether a presenter has "phobic" nervousness or "normal" nervousness before suggestion of a remedy can be given.

977.   Potvin, Janet H.   "The Simulated Professional Meeting: A Context for Teaching Oral Presentation in the Technical Communication Course."   Journal of Technical Writing and Communication 14 (1984):   59-68.

Discusses the use of simulated professional meetings to teach technical communication students at The University of Texas at Austin how to prepare for and present effective oral presentations.

978.   Raymond, Richard C.   "Oral Communication and the Recommendation Report."   Technical Writing Teacher 11 (1984):   105-09.

Stresses the importance of speaking and listening skills to technical communicators.   Describes an assignment that consists of modifying a recommendation report to include interviewing professionals, writing report, and presenting oral reports of their findings.

979.   Rubens, Philip M.   "Oral Grading Techniques: An Interactive System for the Technical Writing Classroom." Technical Writing Teacher 10 (1982):   41-44.

Suggests using cassette tapes to record students' oral presentations.   Offers guidelines for grading oral presentations.

980.   Smith, Barbara.   " 'What? I Have to Give a Speech?' Integrating Oral Presentations into a Technical Writing Course."   Technical Writing Teacher 8 (1981):   59-61.

Describes how to incorporate oral presentation assignments into the technical writing classroom.   Includes a sample evaluation sheet for oral presentations.

981. Wyllie, James. "Oral Communications: Survey and
      Suggestions." ABCA Bulletin 43.2 (1980): 14-17.
          States that more attention should be given to the
      teaching or oral presentations. Provides guidelines
      for teaching oral presentations.

11.2 Designing and Delivering
Oral Presentations

982. Abshire, Gary M., and Kenneth L. White. "Techniques
      and Suggestions for Presenters." IEEE--Transactions
      on Professional Communications PC-21 (1978): 97-100.
          Examines techniques for presenting effective oral
      presentations.

983. Adamy, David L. "Technical Talks: They're Not as
      Hard as You Think." IEEE--Transactions on Profes-
      sional Communications PC-23 (1980): 22-25.
          Offers suggestions for preparing effective oral pre-
      sentations.

984. Baird, John E. "How to Overcome Errors in Public
      Speaking." IEEE--Transactions on Professional Com-
      munications PC-24 (1981): 94-98.
          Suggests ways for eliminating errors from oral pre-
      sentations.

985. Bolmer, Joan. "Tips on Talking in Public." IEEE--
      Transactions on Professional Communications PC-25
      (1982): 40-42.
          Gives suggestions for preparing and delivering oral
      presentations.

986. Costello, John. "Jests Can Do Justice to Your Speeches."
      IEEE--Transactions on Professional Communications PC-
      21 (1978): 60-63.
          Explains how humor can be used to improve oral
      presentations.

987. Cox, John L. "Technical Topic, Lay Audience: How
      to Make a Good Presentation." IEEE--Transactions
      on Professional Communications PC-23 (1980): 26-28.
          Suggests ways to effectively present an oral pre-
      sentation on a technical topic to a lay audience. Provides

guidelines for preparing report content and visual aids.

988.   Crow, Porter J.   "How to Change Their Minds."   IEEE
       --Transactions on Professional Communications PC-23
       (1980):   48-49.
           Provides tips for presenting a persuasive oral pre-
       sentation.

989.   Curtis, Ethel I.   "Three Basic Recipes for a Speech."
       IEEE--Transactions on Professional Communications PC-
       23 (1980):   28-31.
           Explains a method for easily designing and present-
       ing an oral report.

990.   Decker, Bert.   "A Speech Is Worth a Thousand (Written)
       Words."   IEEE--Transactions on Professional Communica-
       tions PC-27 (1984):   32-34.
           Explains ways to effectively deliver a speech.

991.   Estrin, Herman A., and Edward J. Monahan.   "Effective
       Oral Presentation of Scientific and Technical Informa-
       tion."   Journal of Technical Writing and Communication
       5 (1975):   187-98.
           Provides guidelines for preparing and presenting
       effective oral presentations.

992.   Floyd, Raymond E.   "Presentation Fundamentals."   IEEE
       --Transactions on Professional Communications PC-23
       (1980):   40-42.
           Presents topics of key concern to anyone preparing
       for an oral presentation.

993.   Garrett, Nancy.   "Let's Hear It for the Audience."
       IEEE--Transactions on Professional Communications
       PC-23 (1980):   7-8.
           Examines the importance of audience awareness when
       designing and presenting speeches.

994.   Haskins, William A.   "A Silent Partner for Public
       Speakers."   IEEE--Transactions on Professional Com-
       munications PC-23 (1980):   38-39.
           Discusses the importance of understanding the role
       that "time" has in oral presentations.

995. Jones, Hilary. "I Believe in Basic English." IEEE--
Transactions on Professional Communications PC-23
(1980): 34-37.
Shows how use of language that is familiar to the
speaker will reduce fear of giving oral presentations.

996. Lazaro, Timothy R. "Effective Communication of Technical Information to a Nontechnical Group." Journal
of Technical Writing and Communication 7 (1977): 295-
301.
Suggests ways to improve oral presentation of technical information to a nontechnical audience.

997. Linver, Sandy. "The Importance of the Spoken Image." IEEE--Transactions on Professional Communications PC-23 (1980): 8-10.
Examines ways that a speaker can create a desired
image of himself/herself in the minds of an audience.

998. Lubetkin, Maurice. "How to Speak Up ... or Down."
IEEE--Transactions on Professional Communications
PC-23 (1980): 43-44.
Explains how voice control can be used effectively
when giving oral presentations.

999. Miller, Joseph B., and William F. O'Hearn. "Crucial
Decisions for Technical Speakers." IEEE--Transactions
on Professional Communications PC-23 (1980): 14-17.
Discusses the importance of defining the purpose
and analyzing the audience when preparing for a
speech.

1000. Montalbo, Thomas. "Six Steps to the Lectern." IEEE--
Transactions on Professional Communications PC-23
(1980): 5-6.
Discusses six topics of concern for those preparing
oral presentations.

1001. Perry, Robert E. "Audience Requirements for Technical Speakers." IEEE--Transactions on Professional
Communications PC-21 (1978): 91-96.
Discusses audience requirements to help guide technical speakers as they prepare their presentations.

1002. Ratliff, Gerald Lee. "Performance Guide for Oral

Communication." IEEE--Transactions on Professional
Communications PC-23 (1980):   11-14.
   Presents guidelines for preparing and giving
speeches.

1003.   Sawyer, Thomas M. "Preparing and Delivering an
        Oral Presentation." Technical Communication 26.1
        (1979):   4-7.
           Provides a procedure for preparing and delivering
        oral presentations.

1004.   Schoen, Robert. "Let's Give Better Scientific and
        Technical Talks." IEEE--Transactions on Professional
        Communications PC-23 (1980):   120-22.
           Discusses effective use of conclusions in oral pre-
        sentations.

1005.   Williams, Donald E.   "Idea Clarification: A Matter of
        Predesign." IEEE--Transactions on Professional Com-
        munications PC-23 (1980):   32-34.
           Describes ways that a speaker can make content
        easier to understand.

1006.   Wolff, Michael F.   "When You Have to Get Up and
        Speak." IEEE--Transactions on Professional Communi-
        cations PC-25 (1982):   202-03.
           Provides tips for giving good speeches.

11.3 Comparing Oral and Written
Communication

1007.   Darian, Steven. "Using Spoken Language Features
        to Improve Business and Technical Writing." ABCA
        Bulletin 45.3 (1982):   25-30.
           Examines differences between oral and written com-
        munication. Discusses what oral communication has to
        offer the writer.

1008.   Day, Yvonne Lewis. "The Economics of Writing."
        IEEE--Transactions on Professional Communications
        PC-26 (1983):   4-8.
           Examines differences between the spoken and written
        word. Shows how these differences affect writing.

1009.   De Beaugrande, Robert.   "Cognitive Process and Tech-
        nical Writing: Developmental Foundations."   Journal
        of Technical Writing and Communication 12 (1982):
        121-45.
            Examines the influence of speech habits on techni-
        cal writing.

1010.   Eadie, William F., and Michael Z. Sincoff.   "Technical
        Communication in Written and Oral Modes."   Journal
        of Technical Writing and Communication 7 (1977):   205-
        17.
            Discusses the importance of possessing good oral
        and writing skills.   Shows differences and similarities
        that exist between oral and written communication.

1011.   Fitch, Suzanne P.   "A Comparison of the Scientific
        Paper and the Problem/Solution Speech."   Journal of
        Technical Writing and Communication 13 (1983):   29-
        32.
            Shows similarities and differences between presen-
        tation of scientific reports and presentation of problem/
        solution speeches.

1012.   Goldstein, Jone Rymer.   "Integrating Interpersonal
        and Small Group Oral Communication Skills into the
        Technical Writing Course."   International Technical
        Communication Conference.   St. Louis Park, Minne-
        sota, 14-17 May 1980.
            Stresses the importance of integrating both oral
        and written communication into the technical writing
        course.

1013.   Hagge, John.   "Strategies for Verbal Interaction in
        Business Writing."   Conference on CCC.   New York,
        New York, 29-31 March 1984.
            Discusses the similarities between speech and busi-
        ness writing.

1014.   Liggett, Sarah.   "Speaking/Writing Relationships and
        Business Communication."   The Journal of Business
        Communication 22.2 (1985):   47-56.
            Discusses the differences that exist between spoken
        and written language.   Explains that the more a per-
        son understands these differences, the better speaker
        and writer that person tends to be; therefore, teachers

should teach business communication students to recognize these differences. Offers possible teaching
methods.

• 12. GRAPHICS •

12.1 Use in Written Reports

1015. "A Technical Communication Course in Graphics and Audiovisuals." Technical Writing Teacher 7 (1980): 56-59.

Describes how to develop a course in graphics and audiovisuals. Briefly explains course design, objectives, and content. Offers comments concerning student response to the course.

1016. Auger, B. Y. "Visual Aids to Clarity." IEEE--Transaction and Professional Communications PC-21 (1978): 71-76.

Examines types of visual aids and explains how each type can be used.

1017. Barton, Ben F., and Marthalee S. Barton. "Toward a Rhetoric of Visuals for the Computer Era." Technical Writing Teacher 12 (1985): 126-45.

Evaluates the current methods used to teach and discuss the use of visual aids. Finds that what is needed is a "rhetoric of visuals for the computer era" (p. 136). Explains how such a rhetoric can improve instruction on the use of visual aids. Suggests ways that technical communication instructors can improve existing instructional methods.

1018. Bodmer, George R. "Graphic Aids for the Technical Writing Student." The Technical Writing Teacher 11 (1983): 15-20.

Offers a discussion of types of graphics available to writers, such as line drawings, photocopying, templates, etc. Suggests guidelines for choosing the type of graphic appropriate for the writing task. Also,

describes equipment that can be used to produce simple, yet professional looking, graphics.

1019. Bradford, Annette N., and David B. Bradford. "Practical and Empirical Knowledge of Photoillustration:
What Is and Is Not Known." Journal of Technical
Writing and Communication 13 (1983): 259-68.
       Evaluates the effectiveness of using photographs in
conjunction with texts. Offers suggestions for further
research of this topic.

1020. Brown, Maurice A. "Graphic Aids in Reporting Technical Information." Journal of Technical Writing and
Communication 8 (1978): 237-41.
       States that effective communication depends on
"familiarity and simplicity of the communication medium.'
Explains how visual and graphic aids, classified as
either static or dynamic, are used to achieve effective
communication.

1021. Buehler, Mary Fran. "Report Construction: Tables."
IEEE--Transactions on Professional Communications
PC-20 (1977): 29-32.
       Shows how to design effective tables.

1022. Cury, Robert. "Visual/Graphic Aids for the Technical Report." Journal of Technical Writing and Communication 9 (1979): 287-91.
       Explains that visual/graphic aids allow readers to
easily comprehend information presented in technical
documents. Briefly lists and describes uses for some
of the most common visual/graphic aids: grids, tables,
bar charts, flow charts, maps, pie diagrams, drawings,
and sketches.

1023. Darian, Steven. "Using Algorithms, Prose, and Graphics for Presenting Technical Business Information."
ABCA Bulletin 46.4 (1983): 26-29.
       Examines the use of visual elements to enhance
communication.

1024. Filley, Richard D. "Opening the Door to Communication Through Graphics." IEEE--Transactions on Professional Communications PC-25 (1982): 91-94.
       States that graphics allow the human brain to

process information more effectively. Categorizes
graphics as simple, moderately complex, and complex.
Stresses the importance of suiting the graphic to the
audience.

1025. Frye, Robert H. "Artistic Technical Training." IEEE--
       Transactions on Professional Communications PC-24
       (1981): 86-89.
           Discusses the use of cartoons, interpretive sketches,
       and process diagrams to enhance technical documents.

1026. George, James E., and Anders Vinberg. "The Display
       of Engineering and Scientific Data." IEEE--Transac-
       tions on Professional Communications PC-25 (1982):
       95-97.
           Presents eight ways that state-of-the-art computer
       graphics can be used to aid one's understanding of a
       complex answer to a sample engineering problem.

1027. Girill, T. R. "Technical Communication and Art."
       Technical Communication 31.2 (1984): 35.
           Discusses assumptions that underlie the use of art
       in technical communication.

1028. Gould, Christopher, and Kathleen Gilstrap. "Planning
       an On-Campus Workshop in Graphics and Technical
       Writing." CEA Forum 13.3-4 (1985): 21-24.
           Describes a workshop that discusses the use of
       graphics in technical writing. Offers guidelines for
       setting up such a workshop.

1029. Gross, Alan G. "A Primer on Tables and Figures."
       Journal of Technical Writing and Communication 13
       (1983): 33-55.
           Uses "the Euclidean vocabulary of point, line, and
       plane" (p. 34) to analyze the effectiveness of tables
       and figures.

1030. "Guide for Patent Drawings." IEEE--Transactions on
       Professional Communications PC-22 (1979): 109-11.
           Provides a condensed version of the Guide for
       Patent Draftsmen's rules for designing patent draw-
       ings.

1031. Hammet, B. F. "The Eye Sees, But the Mind Perceives."

Journal of Technical Writing and Communication 5
(1975):   131-36.
Discusses the importance of visual literacy and
states that one's depth of perception is determined by
one's experiences.

1032. Hanna, J. S. "Six Starts Toward Better Charts."
Technical Communication 29.3 (1982):   4-8.
Explains how to design effective tables.

1033. Herrstrom, David Sten. "Technical Writing as Mapping
Description onto Diagram: The Graphic Paradigms of
Explanation." Journal of Technical Writing and Com-
munication 14 (1984):   223-40.
Discusses the function of diagrams to facilitate the
purpose of explanation.

1034. Huff, Lloyd, and Richard L. Fusek. "An Introduction
to Holography." Technical Communication 26.4 (1979):
9-11.
Defines holography as a method of recording and
displaying three-dimensional images. Discusses the
principles of holography and describes how holography
is applied to science, commercial uses, art, communication, and data storage.

1035. Levin, Amy K. "'Look, Jane!'--Perception Skills in
the English Classroom." English Journal 74.7 (1985):
46-48.
Examines relationships between visual and verbal
perception. Suggests some exercises that can be used
to improve students' visual perceptions.

1036. MacGregor, A. J. "Selecting the Appropriate Chart."
IEEE--Transactions on Professional Communications
PC-21 (1978):   106-07.
Lists and illustrates eight types of graphs. Guide-
lines for using each of the eight types are given.

1037. Magnan, George A. "Technical Drawings and Illustra-
tions." IEEE--Transactions on Professional Communications PC-20 (1977):   239-45.
Gives a brief history of the use of visual aids in
American industry and describes the various types of
technical literature produced in industry. Offers

general guidelines for finding and choosing visual aids
that are appropriate for users of the documents.

1038. Milroy, R., and E. C. Poulton. "Labeling Graphs for
Improved Reading Speed." IEEE--Transactions on
Professional Communications PC-22 (1979): 30-33.
   Evaluates three methods of labeling graphs. Finds
that "direct labeling of the functions" most improves
reading speed.

1039. Murch, Gerald M. "Using Color Effectively: Designing to Human Specifications." Technical Communication 32.4 (1985): 14-20.
   Examines the use of color as a communication tool.
Provides guidelines for the use of color.

1040. Peterson, Becky K. "Tables and Graphs Inprove
Reader Performance and Reader Reaction." The Journal of Business Communication 20.2 (1983): 47-52.
   Describes a study conducted to determine whether
or not the use of tables and graphs improve readability of a document.

1041. Podell, Edwin J. "Mathematics Must Be Effective in
Technical Communication." IEEE--Transactions on
Professional Communications PC-27 (1984): 97-100.
   Explains how to effectively incorporate mathematical
equations into technical communications.

1042. Rogers, Barbara W. "Audio-Visual Resources in Teaching Technical Writing." Teaching English in the Two-
Year College 8 (1981): 135-36.
   Describes how the author had her class at Nash-
ville Tech explore how drawings and tables communicated. Describes the production of a slide-tape presentation that the author now uses in the classroom
to interest students in communication.

1043. Simcox, William A. "A Design Method for Graphic
Communication." ABCA Bulletin 47.1 (1984): 3-7.
   Describes a method that can be used to design ef-
fective visuals.

1044. Spurgeon, Kristene G. "Teaching Graphics in Technical Communication Classes." Conference on CCC,

Dallas, Texas, 26-28 March 1981.
   Encourages teachers to instruct students in the
use of graphics through studying annual company reports and multiple overlays on the overhead projector
and by visiting professional graphic artists in the community.

1045.   Sussman, David.   "Composing Photographs for Tech-
        nical Journals."   IEEE--Transactions on Professional
        Communications PC-28 (1985):   3-12.
           Suggests ways of producing photographs that are
        visually effective.

1046.   Swaino, Jean M.   "Teaching Illustration Throughout
        the Technical Writing Course."   Technical Writing
        Teacher 12 (1985):   29-31.
           Provides step-by-step instructions for integrating
        the teaching of visual aids throughout the entire tech-
        nical writing course.

1047.   Tebeaux, Elizabeth.   "Using Computer Printouts to
        Teach Analysis and Graphics."   Journal of Technical
        Writing and Communication 11 (1981):   13-22.
           Suggests using computer printouts of data in the
        classroom to teach graphics.   Offers guidelines for
        choosing suitable printouts and suggests some sources.
        Explains how teachers can produce challenging and in-
        structive problems from data sheets.

1048.   Trimble, John W.   "Cartoons Can Add Punch to Your
        Technical Manual."   Technical Communication 23.1
        (1976):   10-11.
           Explains how to use cartoons effectively.

1049.   Vogt, Herbert E.   "You Don't Have to Be an Artist
        to Produce an Illustrated Document."   IEEE--Transac-
        tions on Professional Communications PC-25 (1983):
        108-12.
           Shows how to produce illustrations for documents.

1050.   Wisley, Forrest G., and C. Edward Streeter.   "Toward
        Defining the Function of Visuals Used to Support a
        Verbal Narration."   Educational Technology 25.11
        (1985):   24-26.
           Provides a comprehensive examination of the use of
        visuals to support "verbal narration."

1051. Yates, JoAnne. "Graphs as a Managerial Tool: A Case Study of Du Pont's Use of Graphs in the Early Twentieth Century." The Journal of Business Communication 22.1 (1985): 5-33.
Uses Du Pont as a specific example to detail the historical development of graphs as managerial tools. Explains that as size of businesses, and thus amounts of information handled by businesses, grew in the early twentieth century, managers turned to graphs as means for communicating information and/or analyzing data.

12.2 Uses in Oral Presentations

1052. "Communicating Through Poster Sessions." IEEE--Transactions on Professional Communications PC-22 (1979): 137-40.
Offers suggestions for how to present a poster session at a conference.

1053. Kerfoot, Glenn. "Let Audience See Your Presentation." IEEE--Transactions on Professional Communications PC-23 (1980): 50-51.
Discusses the use of visual aids in oral presentations.

1054. Roberts, Elizabeth A. "Advice for the User of Visual Aids." Technical Communication 26.4 (1979): 15-17.
Presents requirements for effective visual aids. Lists and briefly describes seven types of aids. Provides guidelines for use of visual aids.

1055. Stratton, Charles R., and Edward J. Breidenbach. "Inexpensive Visuals for Oral Presentations." Technical Communication 23.2 (1976): 2-5.
Suggests that hiring a professional graphics person to prepare visuals is best. States that whenever one is limited by a budget, one can use 35mm color slides of typewritten cards arranged on colored burlap backgrounds. Provides step-by-step procedures for producing these slides.

1056. Szoka, Kathryn. "A Guide to Choosing the Right Chart Type." IEEE--Transactions on Professional

Communications PC-25 (1982):   98-101.
    States that appropriate charts should be chosen to
enhance key points given in an oral presentation.
Describes four types of charts.

1057. Davis, Karen, and others. "Three Hours in the Library Now, or Three Weeks in the Library Later." <u>ABCA Bulletin</u> 41.3 (1978): 29-32.
Describes how a teacher can help technical writing students learn about the library facilities that will assist in the preparation of their research reports.

1058. Dragga, Sam. "Technical Writing and Library Research: Pairing Objective." <u>Technical Writing Teacher</u> 12 (1985): 107-10.
Suggests that technical writing teachers link writing assignments to library research so that the technical writing classroom "builds" on students' experiences. Discusses two effective library assignments: "1) writing a comparative analysis of the bibliographies, indexes, and abstracts available" in the students' fields and "2) writing a user's guide for a selected research tool" (p. 108).

1059. Krimm, Bernard, and Stuart Golgoff. "The Information Resources Report: A Method for Integrating Library Instruction with Technical Writing Courses." <u>Technical Writing Teacher</u> 7 (1979): 35-38.
Explains the importance of having students become familiar with library resources in their major fields. Suggests that one way to familiarize students with library resources is to have students "write an evaluative report on the information resources in their particular field or speciality within a field" (p. 35). Also suggests other assignments which can be generated from the resources report.

1060. Minor, Dennis E., and Lowell F. Lynde. "Library Research Material for Technical Writing." <u>Technical</u>

Writing Teacher 3 (1976):   135-43.
　　Provides a list of library reference materials to
help teachers and students of technical writing know
where to search for information on scientific and technical topics.

1061.　Pritchard, Eileen.　"Teaching Science Citation Index
　　　　for a Library Orientation."　Journal of Technical Writ-
　　　　ing and Communication 9 (1979):   297-301.
　　　　　　Provides an effective method for teaching students
　　　　how to use the Science Citation Index.　Includes a
　　　　sample handout that can be used while explaining the
　　　　Index and a sample assignment that students can com-
　　　　plete by actually using the Index.

1062.　Speers, Marthalee A.　"One Good Way to Teach Li-
　　　　brary Research to Business Writing Students."　ABCA
　　　　Bulletin 46.4 (1983):   32-33.
　　　　　　Describes a library report assignment that requires
　　　　students to research a company or organization that
　　　　could be a possible employer.

1063.　Tebeaux, Elizabeth.　"The Importance of Following Up
　　　　Library Instruction."　Journal of Technical Writing
　　　　and Communication 9 (1979):   27-32.
　　　　　　Describes the library instruction procedure used
　　　　to teach students at Texas A&M University how to use
　　　　the library and where to find reference materials with-
　　　　in the library.　Describes the content of and purpose
　　　　of a library report that students must complete as a
　　　　followup to the library instruction.　Finally, explains
　　　　how the library report is used in conjunction with the
　　　　proposal assignment to help students produce formal
　　　　reports that have been more thoroughly researched.

PART III:

NEW HORIZONS

14. COMPUTERS AND TECHNICAL COMMUNICATION •

14.1 Computers in the Classroom

1064. Golen, Steven. "Using a Data Processing Tool in Business Report Writing Class." Journal of Business Education 57.7 (1982): 266-67.
    Describes a method for teaching students how to use a data processor.

1065. Harrington, Henry R., and Richard E. Walton. "The Warnier-Orr Diagram for Designing Essays." Journal of Technical Writing and Communication 14 (1984): 193-201.
    Explains how this design tool for computer programming can be used to help design expository essays.

1066. Johnson, Mildred I., and Karen S. Sterkel. "Computer Text Analysis: For Business, Government, and Classroom Writers." National Forum: Phi Kappa Phi Journal 65.4 (1985): 36-40.
    Describes how a computer program is used to evaluate students' writing in a business writing class offered at Colorado State University.

1067. Kelly, Rebecca S. "Translating Jabberwocky: From Teaching English to Technical Writing." Technical Writing Teacher 9 (1982): 147-49.
    Describes the role that an English teacher can play in bridging the gap between users of computer and writers of computer manuals.

1068. Kelton, Robert W. "Teaching Technical Writing to Computer Science Students (or 'Calm Down, Mr. Chips')" Technical Writing Teacher 11 (1983): 7-14.
    Describes the "life cycle of a data processing

project," showing what written documents accompany
each phase of the project. Explains how the documents written for a data processing project are both
similar to and different from traditional written forms
taught to students in technical writing classrooms.

1069. Kruk, Leonard B. "Word Processing and Its Implica-
       tions for Business Communications Courses." The
       Journal of Business Communication 15.3 (1978): 9-18.
       Discusses the development of word processing and
       analyzes its role in the business world. Then, offers
       suggestions for revising curriculums so that business
       students are prepared to meet the changes that word
       processing has brought to the business office.

1070. Marshall, Stewart. "Computer Assisted Feedback on
       Written Reports." Computers and Education 9 (1985):
       213-19.
       Discusses how a computer program is used to pro-
       vide feedback on students' written reports.

1071. Mikelonis, Victoria M., and Vicki Gervickas. "Using
       Computers in the Technical Writing Classroom: A
       Selected Bibliography (1978-84)." Technical Writing
       Teacher 12 (1985): 161-176.
       Lists and discusses articles according to the follow-
       ing areas: 1) Choosing computer systems, 2) Reviews
       of specific systems, 3) Computers in the technical
       writing classroom, and 4) Automated text-editing and
       publishing, and the professional communicator.

1072. Mullins, Carolyn J., and Thomas W. West. "Word
       Processing for Technical Writers and Teachers." Inter-
       national Technical Communication Society Conference.
       Pittsburgh, Pennsylvania, 20-23 May 1981.
       Describes the computer and word-processing facili-
       ties available at Indiana University.

1073. Oliver, Lawrence J. "The Case Against Computerized
       Analysis of Student Writings." Journal of Technical
       Writing and Communication 15 (1985): 309-22.
       Examines various limitations of computerized analyses
       of writing.

1074. Orth, Michael, and Carl R. V. Brown. "Computer

Generated Rhetorical Simulations for Business and Report Writing Courses." Journal of Technical Writing and Communication 14 (1984): 29-42.

Explains how computer simulations can be used to generate cases that are appropriate for writing assignments.

1075. Rothmel, Steven Zachary. "Taking Stock: Teaching Technical Communication in the Computer Age." Technical Writing Teacher 9 (1981): 16-18.

Examines the role of writers in the computer age. States that computers assist writers in the task of gathering data but do not lessen the writers' job of presenting data to readers in an informative way.

1076. Smith, Edgar Ray. "Integrating Computer Technology in Business Communication Courses: Business Reports and Letters." 31st Annual ABCA Southeast Convention. Hammond, Louisiana, April 5-7, 1984.

Describes how the University of Tennessee has integrated computers into a business writing course.

1077. Stewart, William J. "Using the Computer to Improve Unit Teaching." Journal of Technical Writing and Communication 11 (1981): 265-70.

Describes a Computer-Based Resource Unit developed by Dr. Robert S. Harnak and associates at the State University of New York at Buffalo. Explains how the program is designed to help teachers plan teaching units by providing teachers with suggested objectives, subject matters, activities, materials, and evaluation devices for specified units.

1078. Varner, Iris I., and Patricia Marcum Grogg. "Using Microcomputers in Your Classes Without a Lab." ABCA Bulletin 48.3 (1985): 9-11.

Examines the use of computers in the business communication classroom. Describes how the computer is used in business communication classes at Illinois State University.

1079. Washington, Gene. "Control Structures: Modeling Aspects of the Technical Text." Journal of Technical Writing and Communication 12 (1982): 147-53.

Discusses the use of Sequence, If-Then-Else, and

Do-While structures of computer programming as an invention technique. Explains how these structures help students consider possible ways of organizing information to be presented in texts.

1080. Woolston, Donald C. "Incorporating Microcomputers into Technical Writing Instruction." Engineering Education 75 (1984): 88-90.
        Explains how microcomputers can be used in the technical writing classroom. Discusses the computer hardware, computer software, and assignments used in the author's technical writing course.

14.2 Computers and the Writing Process

1081. Bowman, Joel P. "Plugging In and Turning On to Word Processing." ABCA Bulletin 46.1 (1983): 3-6.
        Discusses various computers, monitors, printers, and word-processing programs that are available.

1082. Catano, James V. "Computer-Based Writing: Navigating the Fluid Text." College Composition and Communication 36 (1985): 309-16.
        Examines the effects of word processing on writing and revision by studying how the use of word processors affects the writing processes of two "accomplished" novelists, Robert Coover and Rosmarie Waldrop. Finds that for these writers, word processing did not stifle creativity and in fact enhanced "the blending of research and writing" (p. 311).

1083. Coleman, Eve B. "Flowcharting as a Prewriting Activity." Computers, Reading and Language Arts 1.3 (1983): 36-38.
        Explains how flowcharting can be used to aid students involved in the writing process.

1084. Grice, Roger A. " 'Cut and Paste' Enters the Computer Age." IEEE--Transactions on Professional Communications PC-27 (1984): 78-81.
        Demonstrates how use of computers can simplify the "cut and paste" process.

1085. Harris, Jeanette. "Student Writers and Word Processing:

A Preliminary Evaluation." College Composition and
Communication 36 (1985): 323-30.
   Describes a study conducted with six students to
determine the effect that word processing has on revising. Finds that the students "made fewer revisions
when they used word processing than when they did
not" (p. 325). Concludes that more research needs
to be done before educators accept the "claims that
word processing can improve our students' writing"
(p. 330).

1086. Horn, William Dennis. "Computer-Assisted Instruction
at Clarkson University." IEEE--Transactions on Professional Communications PC-24 (1984): 197-200.
   Describes various software packages used to facilitate writing instruction at Clarkson University in Pots-
dem, New York.

1087. Lutz, Jean A. "A Study of Revising and Editing at
the Terminal." IEEE--Transactions on Professional
Communications PC-27 (1984): 73-77.
   Describes a study to determine what effect word
processing has on revision and editing.

1088. Martin, Peter. "Ergonomics in Technical Communication." IEEE--Transactions on Professional Communications PC-27 (1984): 62-64.
   Compares technical communicators to user-friendly
computer systems.

1089. Monagle, E. Brette, and O. Jane Allen. "Using the
Computer to Teach Grammar and Mechanics." Technical Communication 30.2 (1983): 9-12.
   Describes a computer program designed to teach
grammar and mechanics. Discusses advantages to using such a program.

1090. Musgrave, Jean F. "Experiments in Computer-Aided
Graphic Expression." IEEE--Transactions on Professional Communications PC-21 (1978): 110-17.
   Examines the possibilities for design of visuals that
computer-aided graphics offer.

1091. Penrose, John M. "Computer Software Review."
ABCA Bulletin 47.3 (1984): 22-24.

Reviews selected software available to assist
writers.

1092. Ross, Peter Burton. "Using the Computer to Teach
Technical Writing." Technical Communication 30.2
(1983): 4-8.
Discusses ways to use the computer to improve
writing.

1093. Rude, Carolyn D. "Word Processing in the Technical
Editing Class." Journal of Technical Writing and Communication 15 (1985): 181-90.
Explains how word processors can be used in a
technical editing class to teach editing skills to students.

1094. Smith, Charles L. "Word Processing and Scientific
Writing in a University Research Group." Technical
Communication 29.3 (1982): 13-16.
Discusses how word processors can be effectively
used to enhance scientific writing.

1095. Soderston, Candace, and Carol German. "A Study of
Analogy and Person in Computer Documentation, Preliminary Report." Presented at the USER-bility Sym-
posium at the International Business Machines Corpora-
tion. Poughkeepsie, New York, July 1984.
Describes a study conducted to determine the effects of the use of analogy and second person on
readers of computer documentation.

1096. Sudol, Ronald A. "Applied Word Processing: Notes
on Authority, Responsibility, and Revision in a Workshop Model." College Composition and Communication
36 (1985): 331-35.
Presents a workshop model for using word processors in the writing classroom. Emphasizes that "our
concern should not be computer applications to writing
but computer applications to writers" (p. 335).

1097. Woelfle, Robert M. "The Impact of Word Processing
on Engineering Communications." IEEE--Transactions
on Professional Communications PC-23 (1980): 159-161.
Describes a computer program designed to teach
grammar and mechanics. Discusses advantages to using such a program.

1098. Wresch, William. "Computers and Composition Instruc-
       tion: An Update." College English 45 (1983): 794-
       99.
           Describes computer programs which "attempt to
       help with the writing process as a whole" (p. 794).
       Comments on research projects that are being con-
       ducted to determine "the effects of word processors
       on student writing" (p. 797).

14.3 Computer Documentation

1099. Annett, Clarence H. "A Structured Approach to Tech-
       nical Project Documentation." Journal of Technical
       Writing and Communication 13 (1983): 299-306.
           Discusses the value of focusing on users and using
       a four-level approach to project documentation.

1100. Antoine, Valerie. "The Software Documentation: A
       New Specialist." Technical Communication 32.3 (1985):
       16-18.
           Examines the characteristics of a good software
       documenter. Discusses various writing tasks that a
       software documenter may undertake.

1101. Asteroff, Janet F. "On Technical Writing and Techni-
       cal Reading." Information Technology and Libraries
       4.1 (1985): 3-8.
           Examines computer documentation, discussing what
       makes computer manuals useful or not. Also provides
       criteria for selecting a computer system.

1102. Bethke, Frederick J. "Measuring the Usability of
       Software Manuals." Technical Communication 30.2
       (1983): 13-16.
           Details how the author developed a tool for measur-
       ing usability of software manuals.

1103. Bradford, David. "The Persona in Microcomputer
       Documentation." IEEE--Transactions on Professional
       Communications PC-27 (1984): 65-68.
           Explains how persona is created and used in micro-
       computer documentation.

1104. Brockmann, R. John, and Rebecca J. McCauley. "The

Computer and the Writer's Craft: Implications for
Teachers." Technical Writing Teacher 11 (1984):
125-35.

Describes a "hierarchy of computer usage in writing" that consists of five levels: true online documents, help screens, pseudo online documents, documents produced by integrated software, and word-
processed documents. Explains how teachers can prepare students to use the computer to produce documents for each level.

1105.   Carliner, Saul. "Help on a Shelf: Developing a
        Customer Service Manual." Technical Communication
        32.3 (1985):   8-11.
        Describes how to develop an effective manual that
        will allow users to diagnose problems with their com-
        puters.

1106.   Gange, Charles, and Amy Lipton. "Word-free Setup
        Instructions: Stepping into the World of Complex
        Products." Technical Communication 31.3 (1984):   17-
        19.
        Describes the use of word-free setup instructions
        for the IBM 5080 Graphics system. Suggests guide-
        lines for producing word-free instructions.

1107.   Grimm, Susan J. "EDP User Documentation: The
        Missing Link." IEEE--Transactions on Professional
        Communications PC-24 (1981):   79-83.
        Discusses the value of user documentation and of-
        fers guidelines for writing good user documentation.

1108.   Henderson, Allan. "The Care and Feeding of the Non-
        Captive Reader." Technical Communication 31.3 (1984):
        5-8.
        Describes changes that are being made in computer
        documentation to meet the changing needs of users.

1109.   Hodgkinson, Richard, and John Hughes. "Developing
        Wordless Instructions: A Case History." IEEE--
        Transactions on Professional Communications PC-25
        (1982):   74-79.
        Details the development of wordless instructions
        for unpacking and setting up the IBM Selectric II.

1110. Little, Raymond Q., and Michael S. Smith. "An Inter-
      active FOMM System." Technical Communication 32.3
      (1985): 11-13.
         Argues for the use of functionally oriented main-
      tenance manuals (FOMMs). Discusses advantages that
      FOMMs offer.

1111. MacIndoe, C. Scott. "An Assessment of Functionally
      Oriented Maintenance Manuals (FOMMs)." Technical
      Communication 32.3 (1985): 7-11.
         Discusses the importance of visuals in making main-
      tenance manuals more readable. Explains how FOMM
      was developed and dispels five myths concerning
      FOMMs.

1112. Meyer, Benjamin D. "The Hardware Conspiracy."
      Technical Communication 32.2 (1985): 12-14.
         Discusses the changing role of computer documenta-
      tion specialists.

1113. Mills, Carol Bergfeld, and Kenneth L. Dye. "Usa-
      bility Testing: User Reviews." Technical Communica-
      tion 32.2 (1985): 40-44.
         Explains the importance of having actual users re-
      view technical documents.

1114. Rubens, Brenda. "Phenomenology, Metaphor, and
      Computer Documentation: A Move Toward a More Self
      Conscious Approach in Technical Writing." Journal of
      Technical Writing and Communication 14 (1984): 19-28.
         Examines the use of more "personalized" writing in
      computer documentation.

1115. Rubens, Philip, and Robert Krull. "Application of Re-
      search on Document Design to Online Displays." Tech-
      nical Communication 32.4 (1985): 29-34.
         Discusses research concerning document design and
      draws from the findings of this research to develop
      guidelines for designing effective online displays.

1116. Thing, Lowell. "What the Well-Dressed Manual Is
      Wearing Today." Technical Communication 31.3 (1984):
      8-12.
         Discusses changes being made in "packaging" of
      computer documentation manuals.

1117. Toms, Roger K. "Writing and Compiling Computer
       User Documentation in a Text Processing Environment."
       Journal of Technical Writing and Communication 13
       (1983):   307-17.
          Examines the importance of computer user documen-
       tation and explains how to develop effective user docu-
       mentation.

14.4 On-Line Computer Instruction

1118. Al-Awar, Janan; Alphonse Chapanis; and Randolph W.
       Ford. "Tutorials for the First-Time Computer User."
       IEEE--Transactions on Professional Communications
       PC-24 (1980):   30-37.
          Offers guidelines for creating effective online tu-
       torials to introduce users to computers.

1119. Blaisdell, F. J. "Historical Development of Computer
       Assisted Instruction." Journal of Technical Writing
       and Communication 8 (1978):   253-68.
          Examines the development of computer assisted in-
       struction.

1120. Bradford, Annette Norris. "Conceptual Differences
       Between the Display Screen and the Printed Page."
       Technical Communication 31.3 (1984):   13-16.
          Discusses changes that must be made when present-
       ing information on a display screen. Offers sugges-
       tions for designing effective display screens.

1121. Rubens, Philip M. "Computer Impact on Teaching
       Technical Communication." Technical Writing Teacher
       10 (1983):   227-31.
          Explains how "technical writing programs could
       benefit from such peripheral computing concerns as
       computer-assisted instruction (CAI), readability pro-
       grams, and language analysis as learning tools" (p.
       227). Examines both advantages and disadvantages
       of using computers in technical communication programs.

15. TELECOMMUNICATIONS AND OTHER • NEW HORIZONS

1122. Andera, Frank. "Important Implications of Letter Mail Automation and Word Processing on Today's Written Communications." ABCA Bulletin 48.3 (1985): 45-49.
Presents suggestions for changing letter-writing procedures in order to accommodate technological changes that are affecting the workplace.

1123. Barkman, Patricia. "The Storyboard Method: A Neglected Aspect of Organizational Communication." ABCA Bulletin 48.3 (1985): 21-23.
Discusses the use of storyboards as an organizational method of communication.

1124. Barton, Ben F., and Marthalee S. Barton. "Communication Models for Computer-Mediated Information Systems." Journal of Technical Writing and Communication 14 (1984): 289-306.
Argues for the importance of examining communication theory and models to improve information-system development.

1125. Duffy, P. R. "Cybernetics." The Journal of Business Communication 21.1 (1984): 33-41.
Describes the development of cybernetics. Examines the impact of cybernetics on information theory and business communication.

1126. Gould, John W., P. Thomas McGuire, and Chan Tsang Sing. "Adequacy of Hong Kong-California Business Communication Methods." The Journal of Business Communication 20.1 (1983): 330-40.
Examines seven methods of communication used by

companies in California and Hong Kong.   Discusses
the advantages and disadvantages of each methods.

1127. Lord, W. J., Jr.   "Information Analysis."   The Jour-
       nal of Business Communication 21.1 (1984):   7-18.
            Discusses the benefits of information analysis.   Ex-
       amines research conducted in this area.   Offers sug-
       gestions for the teaching of business communication
       students.

1128. Lynch, David.   "MIS: Conceptual Framework, Criti-
       cisms, and Major Requirements for Success."   The
       Journal of Business Communication 21.1 (1984):   19-31.
            Evaluates the role of management information sys-
       tems in manager's decision making processes.

1129. Mitchell, Robert B.; Marian C. Crawford; and R. Burt
       Madden.   "An Investigation of the Impact of Electronic
       Communication Systems on Organizational Communica-
       tion Patterns."   The Journal of Business Communica-
       tion 22.4 (1985):   9-16.
            Describes a study made on the effects that elec-
       tronic communication systems have on communication
       within organizations.

1130. Penrose, John.   "Telecommunications, Teleconferencing,
       and Business Communications."   The Journal of Busi-
       ness Communication 21.1 (1984):   93-111.
            Examines the development of telecommunications,
       and, more specifically, teleconferencing.   Describes
       the effect that teleconferencing is having and will
       have on business communications.

1131. Rosetti, Daniel K., and Theodore J. Surynt.   "Video
       Teleconferencing and Performance."   The Journal of
       Business Communication 22.4 (1985):   25-31.
            Describes a study conducted to compare the ef-
       fectiveness of "face-to-face meetings" versus video
       teleconferences.   Finds that video teleconferencing may
       be more advantageous.

1132. Smeltzer, Larry, and Steven Golen.   "Transmission
       and Retrieval of Information: Statements and Hypothe-
       ses of Research."   The Journal of Business Communi-
       cation 21.1 (1984):   81-91.

Discusses technological advances that are associated with concerns of transmission and retrieval of information.   Examines research "hypotheses" that concern information transmission and retrieval.

1133.   Steele, Louise W.   "The Least You Should Know About Office Machines (A Talk to Students)."   ABCA Bulletin 47.3 (1984):   11-17.
Describes various pieces of office equipment that students should know are available.

• 16. SELECTED BOOKS •

1134. Alvarez, Joseph A. The Elements of Technical Writing. New York: Harcourt, Brace, Jovanovich, 1980.

1135. Anderson, Paul, and John Brockman, eds. New Essays in Technical and Scientific Communications: Theory, Research, and Practice. Farmingdale, NY: Baywood Publishing Co., Inc., 1983.

1136. Anderson, W. Steve, and R. Don Cox. The Technical Reader: Readings in Technical Communication. New York: Holt, Rinehart and Winston, Inc., 1984.

1137. Arthur, Richard. Engineer's Guide to Better Communication. Glenview, IL: Scott, Foresman and Co., 1984.

1138. Barnett, Marva T. Writing for Technicians. 2nd ed. New York: Delmar Publishers, 1981.

1139. Bjelland, Harley. Technical Writing: The Easy Way. Lomita, CA: Norway Books, 1981.

1140. Bricq, Ron S. Technically--Write!: Communicating in a Technological Era. 2nd ed. Englewood Cliffs, NJ: Prentice-Hall, 1981.

1141. Brunner, Ingrid; J. C. Mathes; and Dwight W. Stevenson. The Technician As Writer: Preparing Technical Reports. 1st ed. Indianapolis: Bobbs-Merrill Educational Pub., 1980.

1142. Craig, Harkins, and Daniel L. Plung, eds. A Guide for Writing Better Technical Papers. New York: IEEE Press, 1982.

1143. Day, Robert A. How to Write and Publish a Scientific Paper. 2nd ed. Philadelphia: ISI Press, 1983.

1144. Drobnic, Karl, et al. Sci Tech: Reading & Writing The English of Science and Technology. Culver City, CA: English Language Services, 1981.

1145. Eisenberg, Anne. Effective Technical Communication. New York: McGraw-Hill, 1982.

1146. Evans, John. Beginner's Guide to Technical Writing. Stoneham, MA: Focal Press, 1983.

1147. Fabish, Susan. On-The-Job Technical Writing. Hinsdale, IL: Continuing Education Systems, Inc., 1984.

1148. Goswami, Dixie; Janice C. Redish; Daniel B. Felker; and Alan Siegel. Writing in the Professions. Washington, DC: American Institutes for Research, 1981.

1149. Gould, Jay, and Wayne Losano. Opportunities in Technical Communication. Lincolnwood, IL: National Textbook Co., 1983.

1150. Hart, Stuart L., et al., eds. Improving Impact Assessment: Increasing the Relevance and Utilization of Technical and Scientific Information. Boulder, CO: Westview Press, 1984.

1151. Harty, Kevin J. Strategies for Business and Technical Writing. New York: Harcourt Brace Jovanovich, 1980.

1152. Hirschhorn, Howard H. Writing for Science, Industry and Technology. New York: D. Van Nostrand, 1980.

1153. Hoover, Hardy. Essentials for the Scientific and Technical Writer. New York: Dover Publishing Inc., 1981.

1154. Houp, Kenneth W., and Thomas W. Pearsall. Reporting Technical Information. 4th ed. New York: Macmillan, 1980.

1155. Jones, W. Paul. Writing Scientific Papers and Reports. 8th ed. Revised by Michael L. Keene. Dubuque, IA: W. C. Brown Co., 1981.

1156. Jordan, Michael P. Fundamentals of Technical Descrip-
       tion. Melbourne, FL: Robert E. Krieger Publishing
       Co., Inc., 1984.

1157. Journet, Debra, and Julie L. Kling. Readings for
       Technical Writers. Glenview IL: Scott, Foresman and
       Co., 1983.

1158. Larrimore, Randall, and Sara N. Drew. Writing Skills
       for Technical Students. Englewood Cliffs, NJ: Pren-
       tice-Hall, 1982.

1159. Lawrence, Nelda R., and Elizabeth Tebeaux. Writing
       Communications in Business and Industry. 3rd ed.
       Englewood Cliffs, NJ: Prentice-Hall, 1981.

1160. Leonard, David C., and Peter J. McGuire. Readings
       in Technical Writing. New York: Macmillan Publish-
       ing Co., Inc., 1983.

1161. Losana, Wayne, et al. Manual for Technical Writing
       and Business Communication. Dubuque, IA: Kendall/
       Hunt Publishing Co., 1983.

1162. Mackenzie, R. N., et al. Technical Writing: Forms
       and Formats. Dubuque, IA: Kendall/Hunt Publishing
       Co., 1983.

1163. Maimon, Elaine, et al. Writing in the Arts and Sci-
       ences. Englewood Cliffs, NJ: Winthrop Publishing
       Co., 1981.

1164. Markel, Michael H. Technical Writing: Situations and
       Strategies. New York: Saint Martin's Press, Inc.,
       1984.

1165. Marty, Kevin. Strategies for Business and Technical
       Writing. New York: Harcourt Brace Jovanovich, 1980.

1166. McGehee, M. The Complete Guide to Writing Software
       User Manuals. Cincinnati, OH: Writers Digest Books,
       1984.

1167. McQuaid, Robert W. The Craft of Writing Technical
       Manuals. San Diego, CA: Robert W. McQuaid, 1983.

1168. Mehaffy, Robert E. Writing for the Real World. Glen-
       view, IL: Scott, Foresman and Company, 1980.

1169. Messer, Donald K. Style in Technical Writing: A
       Text Workbook. Blenview, IL: Scott, Foresman and
       Company, 1982.

1170. Miles, James L., et al. Technical Writing: Principles
       and Practice. Chicago: Science Research Association,
       Inc., 1982.

1171. Miller, Gary M. Modern Electronics Communication.
       2nd ed. Englewood Cliffs, NJ: Prentice-Hall, Inc.,
       1983.

1172. Olsen, Leslie, and Thomas Ruchin. Principles of
       Communication for Science and Technology. New York:
       McGraw-Hill Book Co., 1983.

1173. Olson, Gary, and James DeGeorge. Style in Technical
       Writing. New York: Random House, Inc., 1983.

1174. Pickett, Nell A., and Ann A. Laster. Technical Eng-
       lish: Writing, Reading, and Speaking. New York:
       Harper and Row Publishers, Inc., 1984.

1175. Puzman, Josef, and Radoslav Porizek. Communication
       Control in Computer Networks. New York: John Wiley
       and Sons, Inc., 1981.

1176. Rogers, Raymond A. How to Report Research and
       Development Findings to Management. Revised ed.
       New York: Pilot Books, 1984.

1177. Rook, Fern. Slaying the English Jargon. Washington,
       DC: Society for Technical Communication, 1983.

1178. Schmidt, Steven. Creating the Technical Report.
       Englewood Cliffs, NJ: Prentice-Hall, Inc., 1983.

1179. Sherman, Theodore A., and Simon Johnson. Modern
       Technical Writing. 4th ed. Englewood Cliffs, NJ:
       Prentice-Hall, Inc., 1983.

1180. Sides, Charles H. How to Write Papers and Reports

About Computer Technology. Philadelphia: ISI Press, 1984.

1181. Smock, Winston. Technical Writing for Beginners. Englewood Cliffs, NJ: Prentice-Hall, Inc., 1984.

1182. Souther, James W., and Myron L. White. Technical Report Writing. 2nd ed. Melbourne, FL: Robert E. Krieger Publishing Co., Inc., 1984.

1183. Sparrow, W. Keats, and Nell Ann Pickett, eds. Technical and Business Communication in Two-Year Programs. Urbana, IL: National Council of Teachers of English, 1983.

1184. Stanley, William. Electronic Communication Systems. Reston, VA: Reston Publishing Co., Inc., 1982.

1185. Stevenson, Dwight W., et al. Courses, Components, and Exercises in Technical Communication. Urbana, IL: National Council of Teachers of English, 1980.

1186. Stratton, Charles R. Technical Writing: Process and Product. New York: Holt, Rinehart and Winston, Inc., 1984.

1187. Turner, Rufus P. Grammar Review for Technical Writers. Melbourne, FL: Robert E. Krieger Publishing Co., Inc., 1981.

1188. Warsin, Thomas L. Teacher's Resources for Technical Writing: Purpose, Process, and Form. Belmont, CA: Wadsworth Publishing Company, 1985.

1189. Weiss, Edmond H. The Writing System for Engineers and Scientists. Englewood Cliffs, NJ: Prentice-Hall, Inc., 1982.

1190. Wisman, Herman M. Basic Technical Writing. 4th ed. Columbus, OH: C. E. Merrill Publishing Co., 1980.

● 17. SELECTED BIBLIOGRAPHIES ●

1191. Alred, Gerald J.; and Diana C. Reep; and Mohan R.
Limaye. Business and Technical Writing: An An-
notated Bibliography of Books, 1880-1980. Metuchen,
NJ: The Scarecrow Press, Inc., 1981.

1192. Balachandran, Sarojini. Technical Writing: A Biblio-
graphy. Urbana, IL: A Joint Publication of American
Business Communication Association and Society for
Technical Communication, 1977.

1193. Bankston, Dorothy, and others. "1975 Bibliography
of Technical Writing." Technical Writing Teacher 4
(1976): 32-43.

1194. Barth, Rodney J. "ERIC/RSC Report: An Annotated
Bibliography in the Communication Arts." English
Journal, 67, No. 1 (1978): 104-08.

1195. Book, Virginia Alm, and others. "1980 Bibliography
of Technical Writing." Technical Writing Teacher 9
(1981): 35-47.

1196. _____. "1976 Bibliography of Technical Writing."
Technical Writing Teacher, 5 (1977): 26-35.

1197. Bowman, Mary Ann. "Books on Business Writing and
Technical Writing in the University of Illinois Library."
The Journal of Business Communication 12.2 (1975):
33-68.

1198. Carlson, Helen V.; Ruth Hersch Mayo; Theresa Am-
mannito Philler; and Douglas J. Schmidt. An Annotated
Bibliography on Technical Writing, Editing, Graphics
and Publishing, 1966-1980. Washington, DC: Society
for Technical Communication, 1983.

1199. Cory, W. A., Jr., comp. "Comprehensive Bibliography
       on Scientific and Technical Writing." Journal of Col-
       lege Science Teaching 11 (1982): 351-55.

1200. Donovan, Robert B., ed. Technical Writing Texts
       for Secondary Schools, Two-Year Colleges, and Four-
       Year Colleges. Washington, DC: National Council of
       Teachers of English Committee on Technical and Sci-
       entific Writing, 1977.

1201. Edlefsen, Karen A. "An Annotated Bibliography of
       the Journal of Technical Writing and Communication,
       1971-1977." Journal of Technical Writing and Com-
       munication 9 (1979): 69-94.

1202. Fearing, Bertie E., and Thomas M. Sawyer. "Speech
       for Technical Communicators: A Bibliography." IEEE--
       Transactions on Professional Communications PC-21
       (1980): 53-60.

1203. Goldstein, Jone Rymer, and Robert B. Donovan. (Dave
       L. Carson, ed.) A Bibliography of Basic Texts in
       Technical and Scientific Writing. Washington, DC:
       Society for Technical Communication, 1982.

1204. Kogen, Myra, and the CCCC Committee on Technical
       Communication. "Bibliography on Education in Tech-
       nical Writing and Communication, 1978-1983." Techni-
       cal Communication 31.4 (1984): 45-49.

1205. McNutt, Anne S. An Annotated Bibliography for
       Teachers of Technical Writing. Available through ERIC
       Clearinghouse, 1977.

1206. Navarro, Richard. "An Annotated Bibliography of the
       Journal of Technical Writing and Communication 1978-
       1980." Journal of Technical Writing and Communication
       12 (1982): 43-56.

1207. Warren, Thomas L. "Select Bibliography: Help for
       Teachers of Technical Writing." Southeastern Con-
       ference on English in the Two-Year College. Winston-
       Salem, North Carolina, 25-27 February 1982.

1208. _____. "The Passive Voice Verb: A Selected,

Annotated Bibliography--Part I." Journal of Technical Writing and Communication 11 (1981): 271-86.

1209.   . "The Passive Voice Verb: A Selected, Annotated Bibliography--Parts II-V." Journal of Technical Writing and Communication 11 (1981): 373-89.